1995

Urban Innovation

Urban Innovation

Creative Strategies for Turbulent Times

TERRY NICHOLS CLARK
EDITOR

SAGE Publications
International Educational and Professional Publisher
Thousand Oaks London New Delhi

For information address:

SAGE Publications, Inc.
2455 Teller Road
Thousand Oaks, California 91320

SAGE Publications Ltd.
6 Bonhill Street
London EC2A 4PU
United Kingdom

SAGE Publications India Pvt. Ltd.
M-32 Market
Greater Kailash I
New Delhi 110 048 India

Printed in the United States of America

Library of Congress Cataloging-in-Publication Data

Main entry under title:

Urban innovation: creative strategies for turbulent times / edited by
Terry Nichols Clark.
p. cm.
Includes bibliographical references and index.
ISBN 0-8039-3800-4.—ISBN 0-8039-3801-2 (pbk.)
1. Urban policy—United States. 2. Municipal government—United States. 3. Municipal finance—United States. I. Clark, Terry N.
HT123.U74553 1994
307.76′0973—dc20 94-13485

94 95 96 97 98 10 9 8 7 6 5 4 3 2 1

Sage Production Editor: Diana E. Axelsen

Contents

Part Three: Coping in Lean Years: Making Fundamental Policy Changes

Foreword

Terry Nichols Clark

The Fiscal Austerity and Urban Innovation (FAUI) Project has become the most extensive study of local government in the world. In the United States it includes surveys of local officials in the approximately 1,000 municipalities with a population greater than 25,000. In some 38 other countries analogous studies are in progress. Although project costs exceed $15 million, they have been divided among project teams so that some have participated with quite modest investments. Our goal is to document and analyze adoption of innovations by local governments and thus to sharpen the information base of what works, where, and why. The project is unusual if not unique in combining a large-scale sophisticated research effort with decentralized data collection, interpretation, and policy analysis. The project's potential to help cities provide better services at lower costs has heightened interest by public officials. The wide range of survey items makes the data base unique for basic researchers on many related topics. Some data are available to interested researchers via the Interuniversity Consortium for Social and Political Research, Ann Arbor, Michigan. The project remains open to persons interested in participating in different ways, from attending conferences to analyzing the data or publishing in our *Newsletter* and in the annual volume *Research in Urban Policy*.

Background

The Fiscal Austerity and Urban Innovation project emerged in summer 1982. Terry Clark, Richard Bingham, and Brett Hawkins had planned to survey how 62 U.S. cities adapted to austerity. We circulated a memo summarizing the survey and requesting suggestions. The response was overwhelming: people across the United States and in several countries volunteered to survey leaders in their areas, covering their own costs. Participants were initially attracted by the opportunity to compare their cities with others near them. When it seemed clear that we would be getting data from most of the United States, others volunteered to survey remaining states. The result was a network of about 26 U.S. teams using a standard methodology to survey local public officials. The teams pooled their data, and then made the information available to all. The project spread internationally in the same manner.

Although the project emerged quite spontaneously, it built on experiences joining many participants. In the past 15 years quite a few empirical studies have had major impacts on urban policy analysis. Social scientists and policy analysts increasingly use such studies, but data collection costs are so high that individuals cannot find a grant to collect data they might desire. Research funds have progressively declined, yet urban research has increased in sophistication and scale. A collective effort thus offers clear payoffs. This situation, recognized in the late 1970s, was the focus of a conference in 1979 at which 20 persons presented papers which reviewed the best urban policy research to date, outlined central hypotheses, and itemized critical indicators that might be collected in future work. Seven participants (Terry Clark, Ronald Burt, Lorna Ferguson, John Kasarda, David Knoke, Robert Lineberry, and Elinor Ostrom) then extended the ideas from the separate papers in"Urban Policy Analysis: A New Research Agenda." It was published in 1981 together with the separate papers by Sage Publications as *Urban Policy Analysis, Urban Affairs Annual Reviews,* volume 21. Several persons and many topics from *Urban Policy Analysis* found their way into the present project.

The Permanent Community Sample (PCS), a national sample of 62 U.S. cities monitored over 20 years, provides a data base and research experience on which the project was built. Many questionnaire items, and methodologies for studying urban processes, were derived from the PCS. Fresh data have regularly been made publicly available; a small data file, provided with an instruction manual, has been used for teaching at many universities. Several hundred articles and books have used the PCS; the most comprehensive is Clark and Ferguson, *City Money* (1983). Both basic research and public policy questions have been addressed, with regard to such issues as the extent to which cities are fiscally strained, and the solutions they can adopt. These and related issues have been used in the United States in conferences, workshops, and publications involving the U.S. Department of Housing and Urban Development, U.S. Conference of Mayors, International City Management Association, and Municipal Finance Officers Association, as well as state and local affiliates of these groups. International participation has involved similar groups, such as the Germany Association of Cities and many individual local officials.

Project participants came to know each other through professional associations such as the American Political Science Association, International Sociological Association, and European Consortium for Political Research. Meetings in Denver and San Francisco in August 1982 facilitated launching the project. The international component developed via the Committee on Community Research of the International Sociological Association. This committee helped organize a conference in Essen, Germany, in October 1981, which led to three volumes published in English by the German Housing and Urban Development agency: *Applied Urban Research,* edited by G. M. Hellstern, F. Spreer, and H. Wollman. This Essen meeting and a Mexico City meeting in August 1982 helped extend the project to Western Europe and other countries.

Since the project began in 1982, conferences have been held regularly around the world, often with meetings of larger associations, especially the European Consortium for Political Research in spring and the American Political Science Association in summer.

The Survey: The Most Extensive Study to Date of Decision Making and Fiscal Policy in U.S. Cities

The mayor, chair of the city council finance committee, and chief administrative officer or city manager have been surveyed using identical questions in each city in the United States over 25,000, nearly 1,000 cities. Questionnaires were mailed; telephone follow-ups and interviews increased the response rate. Questions dealt with fiscal management strategies the city has used, from a list of 33, such as contracting out, user fees, privatization, across-the-board cuts, reducing workforce through attrition, and deferred maintenance of capital stock. Other items concern revenue forecasting, integrated financial management systems, performance measures, management rights, and sophistication of economic development analyses. Most U.S. data collection was completed in winter and spring 1983. Unlike most studies of local fiscal policy, the project includes items about local leadership and decision-making patterns, such as preferences of the mayor and council members for more, less, or the same spending in 13 functional areas. Other FAUI items are policy preferences, activities, and impact on city government by 20 participants, including employees, business groups, local media, the elderly, city finance staff, and federal and state agencies. Several items come from past studies of local officials and citizens, thus permitting comparisons of results over time. Project participants often share new data for the first year and then make them available to others.

Participants and Coordination

The project board, chaired by William Morris, former mayor of Waukegan, Illinois, includes civic leaders and public officials. Terry Clark is coordinator of the project. Most decisions evolve from collegial discussion. Many participants have 10 to 20 years of experience in working together as former students, collaborators in past studies, and coauthors of many publications. Mark Baldasarre and Lynne Zucker developed the U.S. survey administration procedures. Robert Stein played a lead role in merging U.S. project data from 26 teams with data from the Census and elsewhere. Paul Eberts is coordinating surveys of counties and smaller municipal governments involving more than a dozen other persons in a closely related study. Participants include persons who helped devise the study, collect or analyze data, or participate in conferences and policy implementation activities. Data collection is complete in the United States and most European countries; it is still under way in some others. Resurveys to assess changes have been conducted in several countries and U.S. regions.

FAUI Project Participants in the United States

ARIZONA:	Albert K. Karnig
CALIFORNIA:	Mark Baldasarre, R. Browning, James Danzinger, Roger Kemp, John J. Kirlin, Anthony Pascal, Alan Saltzstein, David Tabb, Herman Turk, Lynne G. Zucker
COLORADO:	Susan Clarke
WASHINGTON, DC:	Jeff Grady, Richard Higgins, Charles H. Levine
FLORIDA:	James Ammons, Lynn Appleton, Thomas Lynch, Susan MacManus
GEORGIA:	Frank Thompson, Cal Clark, Roy Bahl
ILLINOIS:	James L. Chan, Terry Nichols Clark, Burton Ditkowsky, Warren Jones, Lucinda Kasperson, William Morris, Tom Smith, Laura Vertz, Norman Walzer
INDIANA:	David A. Caputo, David Knoke, Michael LaWell, Elinor Ostrom, Roger B. Parks, Ernest Rueter
KANSAS:	Paul Schumaker
LOUISIANA:	W. Bartley Hildreth, Robert Whelan
MAINE:	Lincoln H. Clark, Khi V. Thai
MARYLAND:	John Gist
MASSACHUSETTS:	Dale Rogers Marshall, Peter H. Rossi, James Vanecko
MICHIGAN:	William H. Frey, Bryan Jones, Harold Wolman
MINNESOTA:	Jeffrey Broadbent, Joseph Galaskiewicz
NEW HAMPSHIRE:	Sally Ward
NEW JERSEY:	Jack Rabin, Joanna Regulska, Carl Van Horn
NEW YORK:	Paul Eberts, Esther Fuchs, John Logan, Melvin Mister, Robert Shapiro, Joseph Zimmerman
NORTH CAROLINA:	John Kasarda, Peter Marsden
OHIO:	Steven Brooks, Jesse Marquette, Penny Marquette, William Pammer
OKLAHOMA:	David R. Morgan
OREGON:	Bryan Downes, Kenneth Wong
PENNSYLVANIA:	Patrick Larkey, Rown A. Miranda, Henry Teune, William Van Vliet, Susan Welch
PUERTO RICO:	Carlos Munoz
RHODE ISLAND:	Thomas Anton, Michael Rich

TENNESSEE:	Mike Fitzgerald, William Lyons
TEXAS:	Charles Boswell, Richard Cole, Bryan Jones, Robert Stein, Del Tabel, Robert Lineberry
VIRGINIA:	Robert DeVoursney, Pat Edwards, Timothy O'Rourke
WASHINGTON:	Betty Jane Narver
WISCONSIN:	Lynne-Louise Bernier, Richard Bingham, Brett Hawkins, Robert A. Magill
WYOMING:	Oliver Walter

FAUI Teams Outside the United States

The non-U.S. participants are among the leading urban analysts in their respective countries. In several cases they direct major monitoring studies with multiyear budgets including collection of data directly comparable to those in the United States. Gerd-Michael Hellstern, University of Berlin, initially coordinated the European teams. Harald Baldersheim provided much help with Western Europe and led the Local Democracy and Innovation project in Hungary, Poland, Slovakia, and the Czech Republic. Ed Prantilla coordinated the project on six Asian countries. The survey items are being adapted to different national circumstances, retaining the basic items wherever possible to permit cross-national comparisons.

AUSTRALIA:	John Robbins
ARGENTINA:	Martha Landa
AUSTRIA:	H. Bauer
BELGIUM:	Dr. Strassen, Marcel Hotterbeex, Catherine Vigneron, Johan Ackaert
BULGARIA:	N. Grigorov, O. Panov
CANADA:	Andrew S. Harvey, Caroline Andrews, Dan Chekki, Jacques Levilee, James Lightbody, Mary Lynch
CHINA:	Min Zhou, Yun-Ji Qian
CZECH REPUBLIC:	Michael Illnern, Jiri Patocka
DENMARK:	Carl-Johan Skovsgaard, Finn Bruun, Poul-Erik Mouritzen, Kurt Houlberg Nielsen
FIJI:	H. M. Gunasekera

FINLAND:	Ari Ylonen, Risto Harisalo
FRANCE:	Richard Balme, Jean-Yves Nevers, Jeanne Becquart-Leclercq, P. Kukawka, T. Schmitt, Vincent Hoffmann-Martinot
GERMANY:	B. Hamm, D. H. Mading, Gerd-Michael Hellstern, Oscar Gabriel, Volker Kunz, Frank Brettschneider
GREAT BRITAIN:	A. Norton, P. M. Jackson, Michael Goldsmith, James Chandler. George Boyne, Bryan Jacobs
GREECE:	Elias Katsoulis, Elisavet Demiri
HONG KONG:	P. B. Harris
HUNGARY:	G. Eger, Pateri Gabor
INDONESIA:	Hatomi, Jonker Tamba
IRELAND:	Carmel Coyle
ISRAEL:	Daniel Elazar
ITALY:	Guido Martinotti, Enrico Ercole
JAPAN:	Hachiro Nakamura, Nobusato Kitaoji, Yoshiaki Kobayashi
KENYA:	Daniel Bourmund
NETHERLANDS:	A.M.J. Kreukels, Tejo Spit
NIGERIA:	Dele Olowu, Ladipo Adamolekun
NORWAY:	Harald Baldersheim, Helge O. Larsen, Johnny Holbeck, Sissel Hovik, Kari Hesselberg, Nils Aarsaether, Solbjorg Sorensen, Synnove Jenssen, Lawrence Rose, Per Arnt Pettersen
PHILIPPINES:	Ramon C. Bacani, Ed Prantilla
POLAND:	Gregory Gorzelak, J. Regulski, Z. Dziembowski, Pawel Swianewicz, Andrzej Kowalczyk, Jerzy Bartowski, Leszek Porebski, Ewa Jurczynska
PORTUGAL:	J. P. Martins Barata, Maria Carla Mendes, Juan Mozzicafreddo
REPUBLIC OF KOREA:	Choong Yong Ahn
SENEGAL:	Abdul Aziz Dia
SLOVAKIA:	Gejza Blaas, Miroslav Hettes
SPAIN:	Cesar E. Diaz, Teresa Rojo
SWEDEN:	Hakon Magnusson, Lars Stromberg, Cecilia Bokenstrand, Jon Pierre, Ingeman Elander, P. O. Norell
SWITZERLAND:	A. Rossi, Claude Jeanrenaud, Erwin Zimmerman, Hans Geser, Erwin Ruegg

TAIWAN:	Fang Wang
TURKEY:	U. Ergudor, Ayse Gunes-Ayata
YUGOSLAVIA/ SLOVENIA:	Peter Jambrek

Participation in the project is open and teams continue to join, especially those from outside the United States, who learn of the project and find ways to merge it with their own activities. Austerity links the less affluent countries of the world with others who can learn from them.

Research Foci

Project participants are free to analyze the data as they like, but past work indicates the range of concerns likely to be addressed. The seven-author statement "Urban Policy Analysis: A New Research Agenda" (cited above) outlines several dozen specific hypotheses. Many specific illustrations appear in project publications such as the six volumes of *Research in Urban Policy* (JAI Press) completed to date, *Urban Innovations as Response to Urban Fiscal Strain* (1985), edited by Terry Clark, Gerd-Michael Hellstern, and Guido Martinotti; and several country-specific reports, including William J. Pammer's (1989) *Managing Fiscal Strain in Major American Cities.* Over 200 papers have been presented at project conferences or published separately, as listed in the *Newsletter*. Some general themes follow.

Innovative Strategies Can Be Isolated and Documented

Showcase cities are valuable to demonstrate that new and creative policies can work. Local officials listen more seriously to other local officials showing them how something works than they do to academicians, consultants, or national government officials. Specific cases are essential to persuade. But because local officials seldom publicize their innovations, an outside data-collection effort can bring significant innovations to more general attention. Questions: What are the strategies that city governments have developed to confront fiscal austerity? How do strategies cluster with one another? Are some more likely to follow others as a function of fiscal austerity? Case studies of individual cities detail strategies identified in the survey.

Local Governments That Do and Do Not Innovate Can Be Identified: Political Feasibility Can Be Clarified

One can learn from both failure and success. Local officials often suggest that fiscal management strategies like contracting out, volunteers, and privatization are"politically infeasible"—that they may work in Phoenix, but not in Stockholm. Yet why not—specifically? Many factors are hypothesized, and some studied, but much past work to date is unclear concerning how to make such programs more palatable. The project is distinctive in probing adoption of innovations, tracing diffusion strategies, and sorting out effects of interrelated variables. The project also probes interrelations of strategies with changes in revenues and spending.

National Urban Policy Issues

In several countries, and especially the United States, reductions in national government funding for local programs compound fiscal austerity for cities. How are cities of different sorts weathering this development? Cities are undergoing some of the most dramatic changes in decades. When city officials come together in their own associations or testify on problems to the media and their national governments, they can pinpoint city-specific problems. But they have difficulty specifying how widely problems and solutions are shared across regions or countries. The project can contribute to these national urban policy discussions by monitoring local policies and assessing the distinctiveness of national patterns. The project summarizes nationally fiscal strain indicators of the sort computed for smaller samples of cities. Types of retrenchment strategies are being assessed. Effects of national program changes are being investigated, such as stimulation-substitution issues. A several-hundred-page report of key national trends in 12 countries has been published by Poul-Erik Mouritzen and Kurt Nielsen (1988), *Handbook of Urban Fiscal Data.*

Conclusion

The project is such a huge undertaking that initial participants doubted its feasibility. It was not planned in advance, but evolved spontaneously as common concerns were recognized. It is a product of distinct austerity in research funding, illustrating concretely that policy analysts can innovate in how they work together. But most of all, it is driven by the dramatic changes in cities around the world and a concern to understand them so that cities can better adapt to pressures they face.

References

Clark, T. N. and L. C. Ferguson. 1983. *City Money: Political Processes, Fiscal Strain and Retrenchment.* New York: Columbia University Press.

———, G. M. Hellstern, and G. Martinotti. 1985. *Urban Innovations as Response to Urban Fiscal Strain.* Berlin: Verlag Europaeische Perspektiven.

Hellstern, G. M., F. Spreer, and H. Wollmann. 1982. *Applied Urban Research.* Bonn, Germany: Bundesforschungsanstalt für Landeskunde und Raumordnung.

Mouritzen, P.-E. and K. Nielsen. 1988. *Handbook of Urban Fiscal Data.* Odense, Denmark: University of Odense.

Pammer, W. J. 1989. *Managing Fiscal Strain in Major American Cities.* New York: Greenwood.

Acknowledgments

Thanks first to the wonderful colleagues with whom we launched the Fiscal Austerity and Urban Innovation project in 1982. Together we surveyed 1,030 U.S. cities over 25,000, and many more internationally. In the past decade, we have helped one another refine ideas in conferences, add data, and complete some 200 papers together—on which this book builds in various ways. The citations recognize many contributions; others are in FAUI *Newsletters* (especially Number 21, June 1992).

Approaches to power and organizational innovation were improved in joint courses and discussions with Chicago colleagues Joseph Ben-David, Charles Bidwell, Peter Blau, James Coleman, Edward Laumann, Paul Peterson, and Edward Shils. Political culture approaches were refined through exchanges with Aaron Wildavsky, Seymour Martin Lipset, Ronald Inglehart, and Daniel Elazar.

Students, postdoctoral fellows, and research assistants, mainly at the University of Chicago, have helped us pursue many avenues. Starting years ago merging FAUI files and creating indexes, Rowan Miranda shows in three important chapters below how mastering specifics can translate into creative research. Margaret Burg and Martha Landa helped with the initial analyses of political leadership. The chapter by Cal Clark and Oliver Walter was successively updated by Edgar Bueno, Daniel Crane, Ed Harper, and Ziad Munson. Eric Fong and Yun-Ji Qian painstakingly extracted and documented hundreds of variables from 50 Census tapes. Other prepared memos on many topics: Jamie Struck on intergovernmental relations, Ted Manley on black mayors, Mark Gromala and Michael Rempel on political culture. Others helped on everything else: Jerzy Bartowski, Pawel Swianiewicz, Dennis Merritt, Chris Miles, Edward Vytlacil, Douglas Huffer, Joie Bentrez, James Hoppenworth, Lincoln Quillian, Daniel Crane, and Tiran Kiremidjian.

The FAUI project is remarkable in doing so much with so little funding: the secret is generous contributions of time and resources from many, many persons. We created a nonprofit organization to pursue several overlapping activities, Urban Innovation Analysis, Inc., the board members of which have been unusually wise and helpful, especially Sydney Stein, Jr., Ferdinand Kramer, Marshall Holleb, Melvin Mister, and Mayor William Morris, Chairman. Local officials in Urban Innovation in Illinois brought street smarts, especially Bill Stafford, Charles Schwartz, Mayor Lucinda Kasperson, and Ernest Barefield—who invited me to train for a marathon with him (Sundays from 4 to 8 a.m.) along the Chicago Lakefront while he was chief of staff for Mayor Harold Washington. Professional groups including the U.S. Conference of Mayors, the national and Illinois Government Finance Officers Associations, and others in Illinois contributed staff time, postage, photocopying, space, and good ideas. Melissa Pappas, Robert Eppley, and Ernest Reuter helped translate these into tangible aid to city officials. James Chan, Ron Picur, and I worked together on several city consulting projects. Burton Ditkowsky officiated as a unique sort of godfather. Lorna Ferguson was always there.

Financial aid came from the U.S. Department of Housing and Urban Development, The Ford Foundation, The Joyce Foundation, The Charles F. and Lavinia Schwartz Foundation, the Illinois Department of Commerce and Community Affairs, the City of Chicago, and generous individuals.

Many thanks to all.

Terry Nichols Clark

1

Introduction

Turbulence and Innovation in Urban America

Terry Nichols Clark

Background

This volume asks how cities innovate. How and why do city governments adopt new policies? Our main focus is U.S. cities, but a few international comparisons are introduced to highlight the U.S. case.[1] We focus on the last two decades—a period of dramatic socioeconomic and political change that social scientists have minimally incorporated into their theories. We advance social science theory by identifying some key social changes, by showing how these demand adjustments in past assumptions, and by building new propositions to incorporate them. The turbulence of the last two decades is particularly critical in reshaping our analyses of how governments work. We have seen:

- The end of more than a century of government growth, manifest especially in cutbacks of grants from national governments to cities—U.S. government grants dropped by more than half from 1977 to the early 1990s.
- A breakdown of traditional distinctions between political parties and disattachment of voters from all parties—illustrated by a more than doubling of Independent voters, as high as 33 percent in some surveys.

- A drop by half in voter turnout in elections from the 1960s to 1980s, but a doubling of membership in certain organized groups (notably ecology groups).
- Taxpayer revolts, like Proposition 13 in California.
- More educated, independent, skeptical, and demanding citizens, dissatisfied with traditional service delivery modes.
- New, talented leaders who develop creative forms of service delivery and distinctive general leadership patterns.

These are just a few examples of the turbulence challenging standard operating procedures of city governments. Yet every change breeds resistance; the new often adds to, instead of replacing, the old. City councils and administrators add some of the new without abandoning the old, struggling to respond to new pressures in ways that seem "reasonable" and "feasible."

How then do cities innovate? We provide two answers. First is a synthesis of work by ourselves and others during the past decade, with examples of innovations that actually work in specific cities. The clearest summary of these points for the practical reader is in our concluding Chapter 8, summarized at the end of this chapter.

Our second answer to how cities innovate takes a longer route. We reassess theories of political leadership and government decision making; we develop a new interpretation. We show how leadership is embedded in changing citizen preferences and organized group activities. We identify a new political culture. It defines new rules of the game. Components have been described by specialists in many subfields, but it is far more than the sum of its parts. The New Political Culture is a dramatic break from the past. It integrates many changes indicated above into a political program. These new policy goals and approaches to decision making stress more collegial management and citizen responsiveness. Our general elaboration of the New Political Culture is in Chapter 2; applications are in other chapters, especially Chapter 4, "The Antigrowth Machine." In developing our approach during the past several years, we found many persons skeptical that really fundamental changes are occurring. One of our basic tasks here is to document the extent and nature of changes in city leadership and policy making. We are fortunate in having an unusually rich and diverse set of sources:

- *To provide a sense of specific leaders, cities, and policies:* brief case studies, often generated from consulting with individual cities and from an awards program we developed on Urban Innovation for city governments.

- *To assess the national scope of major trends, and identify unusual cases:* surveys of mayors, council members, and chief administrative officers in every U.S. city with a population greater than 25,000 (1,030 cities); similar surveys in most of Europe, and selected countries in Asia, Latin America, and Africa, as part of the Fiscal Austerity and Urban Innovation Project. Although this volume only briefly introduces international materials to help interpret U.S. developments, our understanding of processes like class conflict and local autonomy in the United States has been sharpened by a decade of work on such issues around the world.
- *To add baseline variables on socioeconomic changes:* highly detailed data from unpublished U.S. Census tapes, merged with our survey data.
- *To compare with past results by others:* we merged surveys such as those from the International City Management Association (very few other major urban surveys have been conducted since the 1970s).
- *To add historical perspective:* past surveys of U.S. cities, especially the Permanent Community Sample of 63 cities monitored by the National Opinion Research Center at the University of Chicago from 1967 onward.

These combined sources provide far more extensive information than available to us or any others in past work on urban politics and local government. Rapid and continued progress in computer hardware and software has helped us to access, analyze, and present these massive data simply and rapidly. A calculation that in the 1960s took weeks of work by 20 assistants could be completed in minutes by the 1990s.

Yet people, as ever, are the key, especially participants in the Fiscal Austerity and Urban Innovation (FAUI) Project, which began in 1982. Across the United States and around the world, we have conducted common surveys and case studies, identifying innovations, specifying where and why they work. These joint efforts have led to a substantially deeper understanding of these issues than was possible from past studies of local government. For the first time ever, comparable data are available for national samples of cities in countries around the world. We can now identify specifics that past writers had to speculate about or not discuss, such as how much impact do political parties have in different countries, or how important are neighborhood or business groups in affecting local government policies?

The international Fiscal Austerity and Urban Innovation surveys also included 33 policy strategies that many cities have used, such as user fees or contracting out. A descriptive overview of the strategies is in Chapter 8, "Innovations That Work." For a policy-oriented introduction to specific innovations, scan this chapter first.

We do not present a single key finding or solution for urban problems. There isn't one. Our answers are more complex, just as the problems are more complex. Few local officials clearly advocate any single strategy or set of strategies; there is no "widget" that all cities should or can adopt. Instead, what emerges powerfully is that *cities differ* in their problems and appropriate solutions, both inside and outside the United States. But they do not differ randomly. Specifying how and why they differ in broad orientation and specific policies is a major focus of this book and related international work.

We have learned from many local officials. They are our real heroes and heroines. They stand for elected office and win or lose after investing enormous energy and time in return for modest tangible rewards, at least as measured by money. In this period of dramatic transformation in political ideologies, we must refrain from filtering the attitudes and behavior of local officials through our own narrow political perspectives. Studying political ideology does not give license to social scientists to promote their own ideology. Maintaining openness to major transformations is essential to understanding, even if the transformations contradict one's most cherished beliefs. Neither we nor anyone can be completely objective; we are all products of our surroundings. But given the central importance of changing ideologies, critical to extending our understanding is fleshing out alternative ideological perspectives, old and new, and linking them to specific types of citizens, organized groups, cities, and policy outputs.

A policy implication of this perspective is that some city officials find certain classes of strategies more feasible, that is, easier to adopt, than others. Feasibility is more than a choice by one person, it flows from the overall orientation of the city, from its political culture. Clarifying which strategies are more feasible, politically and administratively, and why, is a major concern in the following pages. "Feasibility" for a participant becomes "adoption," "diffusion," or "implementation" for an analyst. Although we build on several past "theories" in this area, most are too narrow, as they stress only one or a few factors and thus hold only in a particular context—which is usually left undefined. This has impaired innovation theories for decades (see Clark 1968). Does "the market" or "entrepreneurial leadership" spur innovation, as certain theories hold? Only in selected contexts, for example cities that differ in their resources, key public participants, and rules of the game (or political culture). For instance, contracting out with private firms to reduce costs may not be politically feasible if unions are extremely powerful. The manager who is highly entrepreneurial and aggressive (as certain policy schools teach) will just get fired in cities where elected officials are jealous of staff. Specifying strategies that "work" in cities with strong unions, for example, and what works elsewhere, must be recognized as a critical part of the theory. The search for a theory providing contextually sharper answers to important policy questions leads us to propositions including contextual relations (for example, contracting in cities with strong versus weak unions). This similarly leads us to focus on political and administrative cultures of different locales. We detail more than has been done in most past work how specific rules of the game and policy preferences (cultures) operate to constrain or facilitate particular policy strategies. Our international work is especially important in indicating what about U.S. cities is unique or shared with cities in other countries.

These analyses contribute to the sociology of knowledge as well as to urban policy analysis in the sense that we show how a theory that works for one context (or city) may be inadequate for another context (or city)—mainly by elaborating the ways that different city cultures define different legitimate rules of the game, for social scientists as well as political leaders and citizens. Comparisons (across cities, time, and national borders) bring out the limits of any one specific theory or policy solution and make clear the importance of theories including context variables as a central explanatory component.

New Leaders and Old Cleavages: What New Rules Are Emerging and Why?

Cities are dramatically shifting their patterns of governance and decision making. Past observers spoke of "power elites," "community power structure," and "business domination." More recently we hear of "runaway pluralism," hyperactive "special interest groups," leadership "vacuums," and find candidates for elective office—from city halls to the White House—campaigning against "bureaucracy" and "overregulation." Yet simultaneously in the 1980s the "homeless" and "underclass" entered our political vocabulary, and new programs were launched to assist them. How do we interpret these ostensibly contradictory changes? Three chapters explore these interrelated aspects of leadership.

We use leadership as a broad concept encompassing the major policy preferences and rules of the game observed by leading elected, appointed, and private sector participants in helping shape major public policy commitments. This is similar to Dahl's "Who governs?" and overlaps with the "regime" of Clarence Stone and others, but differs in not assigning a key role to business leaders (see Chapter 4 on related leadership theories).

Chapter 2, "Race and Class Versus the New Political Culture," shows how two patterns of leadership and political culture are emerging as distinct alternatives in U.S. cities. First is class politics, which dates from the 1930s as a central division between New Deal Republicans and Democrats, allying them respectively with rich or poor, unions or management, and distinct types of organized groups. Whites and Protestants opposed immigrant Catholics and blacks on many issues, mainly material and redistributive, such as pay, workplace conditions, and housing and health. These remain the central issues in many big cities in the 1990s, especially those with large minority populations. Class politics continues more powerfully in cities with more hierarchy, as measured by inequality in income, jobs, and education. These are often attacked as continuing evidence of racism and discrimination. We find indeed that these hierarchies (measured by inequality indexes) generate more active organized groups, and in turn mayors and council members who articulate more vigorous social welfare policies concerned with helping the disadvantaged. Race in U.S. cities works like class and parties in Europe. Our international results show how different hierarchies in different societies crystallize social conflict and drive it ahead politically. Chapter 2 covers more than two dozen measures of hierarchy, including Atkinson Inequality Indexes of income, occupation, education, and national background for each of our U.S. cities. These provide multiple tests of the impact of hierarchy, and results that hold strong in numerous multiple regression models. Whereas the models show that "race" is the visible concern in U.S. cities, these hierarchy measures give it more general analytical meaning, permitting comparisons with hierarchies and (looser) class politics in other countries.

The second pattern of leadership emerged in the 1970s: a new political culture (NPC) that contrasts sharply with class politics. NPC issues include environmentalism, growth management, gay rights, and abortion and in general less concern with the workplace and jobs issues of class politics and more concern with the home, consumption, and lifestyle. Leadership comes less from parties, unions, and ethnic groups in coalitions of high versus low status. Rather, leadership shifts from issue to issue; leaders on abortion are distinct from leaders on environmental issues. Citizens and the media are more active. NPC emerges more fully and forcefully in cities with more highly educated citizens, higher income, and more high-tech service occupations.

One can point to clear cases of each type of political culture. NPC is clear in places like Boca Raton, Florida, with its remarkable environmental concern; or Boulder, Colorado, with tough local regulations on housing and growth; or Austin, Berkeley, and Silicon Valley towns like Palo Alto and Sunnyvale. This culture seemed limited to "college towns" in the 1960s, but subsequently spread much further. On the other hand, Detroit, Chicago, New York, and Los Angeles all had black mayors and seething social problems in the 1980s. They are strong reminders that much of the United States contrasts with NPC. From 1933 to the 1970s, mayors joined national Democratic leaders in the New Deal. But breakdown of the New Deal coalition in Congress ended the clear linkages between cities and national urban policy, and the "substructure" for Democratic class politics of federal grants. In 1978 President Jimmy Carter formally enacted the first National Urban Policy. But paradoxically the 1970s were the very years when the key rules of decision making defining the New Deal—whereby the federal government helped cities via specific programs and substantial funding—were ending. Federal grants to cities dropped precipitously after 1977. Mayors and city residents were granted more fiscal autonomy, a pattern ironically labeled "Fend for Yourself Federalism."

Both patterns—class politics and the New Political Culture—characterize some U.S. cities in the 1990s. Yet if we can distinguish the two conceptually, most cities are still a mix. Leadership is not predetermined by economic or social patterns of any sort. No empirical findings are "deterministic" and there is considerable choice by citizens, group leaders, and elected officials to pursue policies and approaches they themselves generate. This "room for maneuver" is shown by moderate to weak statistical relations between leadership patterns and socioeconomic characteristics of cities.

Although one finds both class politics and the New Political Culture in U.S. society in general, the two are mixed in national politics and most media coverage. The benefit of a local focus is identifying more pure and distinct examples of each process in specific cities. We use both individual sketches of mayors and cities to portray such general patterns and comparative urban data to indicate how unique or representative is each city.

Many urban theorists hold that special interest groups prevail in increasing government growth in general and in directing specific policies. Thus in cities with stronger class and racial differences, groups representing these differences should play key policy roles, pressing the city government for more programs for their constituents. New York is the classic instance, and the New York fiscal crisis of 1975 the classic policy consequence. The polar opposite in this regard was Chicago, whose mighty Democratic party under Mayor Richard J. Daley prevented fiscal crisis. The general lesson, some suggested, was that strong parties can contain aggressive organized groups. Strong leaders, backed by a strong party, can say no to group demands, because their political base is powerful enough to let them say no. Strong parties thus contain spending and government growth. This is the "strong party organization" (SPO) theory that Miranda elaborates.

Rowan A. Miranda's Chapter 3, "Containing Cleavages," makes a distinct contribution in formulating the SPO theory in a set of more abstract yet specific propositions. He uses the Fiscal Austerity and Urban Innovation (FAUI) Project data to test whether the SPO theory holds nationally. The FAUI data permit him to measure party and organized group strength in cities across the United States. This chapter is the first test of the SPO theory using national urban data and a methodology to control effects of confounding variables (such as income, unionization, and racial composition of the city). Miranda finds generally strong and consistent support for the SPO theory: strong parties suppress spending. Of course, "strong" is relative only to other U.S. cities; the result is from comparing U.S. cities on three items tapping the importance of political parties in urban leadership. Miranda also distinguishes SPO from clientelism, which in the form of Irish patronage politics often increased government spending and expanded its functions. Centralization, as in the business leadership of Dallas in past years, worked like an SPO: Dallas business leaders were strong enough that they could ignore group pressures. In Dallas they governed not with a strong party, but via more classic U.S. reform government institutions, combined with a civic/business forum that sponsored candidates for the city council. "Strong centralization" is thus a more general leadership mechanism than a strong party. This is an important point for an era when most hierarchies are withering, business as well as parties. The SPO theory, ceteris paribus, thus warns us that fiscal controls may be harder to impose with no strong leadership.

The SPO theory is consistent with a more abstract proposition: centralization encourages public goods (Clark 1975) if one considers fiscal constraints as a public good. Its converse is: decentralization generates separable (or private) goods. Decentralized leadership in general encourages more specific, tangible projects and services, including clientelism and patronage and higher spending.

Lincoln Quillian (1990) pursued the SPO hypothesis using FAUI data similar to Rowan Miranda's, but included France, Finland, and Japan in addition to the United States. He found minimal support for the theory outside the United States, due it seems to less variation in party importance across localities in other countries and their greater national coherence, organizationally and ideologically. More local differences across U.S. cities permit variations like SPO to exert specific effects. Context—local and national—matters.

Coping With Policy Choices

Chapter 4, "The Antigrowth Machine," asks how new leadership patterns shift policies. Its focus is a new urban policy: growth controls. Such controls are a dramatic reversal of local leaders' traditional efforts to attract jobs and foster growth. Growth is still widely viewed as the "standard," even "necessary" policy for cities, in practice and urban theory. This is a simplified, deterministic view. It is refuted first in brief case studies of San Francisco, Seattle, Los Angeles, Boulder, and Boca Raton, where we show how these early national leaders in growth control adopted stringent policies. We introduce national surveys to document the spread of growth controls across the country.

Besides their current policy importance, growth and antigrowth policies of city governments are a strategic research site for urban theories. Conceptually, growth-limiting movements fundamentally challenge several urban theories that imply that local governments *never* limit growth. Yet this theoretical "conclusion" is obviously refuted by what we see daily: the visible importance of growth management and environmentally sensitive policies. Toward more conceptual clarity, we identify the conceptual roots, and limits, of alternative theoretical perspectives.

Capitalism and impersonal competition among cities are stressed by Paul Peterson and others, who conclude that localities must encourage business development. A more visible hand is sketched in theories by Harvey Molotch, Steven Elkin, and Clarence Stone, whose growth machines and business regimes imply that business wants growth, so government must encourage it. A third theoretical approach, linked to the group theory tradition, stresses social movements, protest, and citizen mobilization as the source of social change—including antigrowth movements. Finally we consider political cultural theories, which stress how rules of the game vary across social groups and cities, implying for instance that growth controls should emerge more in cities with NPCs with younger, more educated and affluent residents. Another political culture proposition suggests that like other social movements, antigrowth activities are encouraged if they must react against hierarchy. The hierarchy here is government leaders committed to growth and unresponsive to citizen and group pressures for change.

Testing these theories, we find that antigrowth *movements* are indeed more likely in cities with NPC-type characteristics and more hierarchical local leaders. Next, several types of *growth management policies* of city governments—from moderate zoning through stringent limits on construction permits and building height—are examined. We find no effects of business leadership on antigrowth movements, or on growth management policies—undermining the growth machine and business regime interpretations that stress business. The most powerful factor explaining adoption of growth management policies is presence of a local antigrowth movement. This supports the group theory tradition. However, in cities where a New Fiscal Populist mayor is in power, then he or she replaces the antigrowth movement in explaining growth-limiting policies. This follows because organized groups no longer need protest outside when one of their own sits in city hall.

We move next to broader policy choices. Why do governments grow or retrench? This question has sparked considerable debate, but little solid evidence. Consider some major hypotheses. William Niskanen argued that budgets grow largely because government bureaucrats seek to expand their agency staffs and perquisites. During the last century, Theda Skocpol and collaborators added historical specifics, but like Niskanen they focus largely on top agency staff, arguing that "the state" must be "brought back in." Why? One neomarxist answer invoked "the fiscal crisis of the state" (e.g., O'Connor) to argue that government growth was necessary to maintain legitimacy and prevent revolution. Habermas, Offe, and others extended this theory, but again supplied minimal evidence about how and why "capitalism" or "legitimacy" should specifically operate to increase government. T. H. Marshall and Daniel Bell stressed expansion of citizenship rights, or "entitlement" in broad historical essays. Richard Rose, Guy Peters, Philippe Schmitter, Harold Wilensky, and others have pursued related ideas with selected European and U.S. data, but again only for national governments in recent decades. Careful empirical work has been done by Peter Flora and associates, detailing national histories of European welfare states since World War II, but only minimally comparing across countries to test more general hypotheses. By focusing on national governments, virtually all these studies have difficulty imputing causality.

This past work has generated many hypotheses, but has failed to disentangle specific causal mechanisms as there are more causes than cases. Sensitive judgment is thus used rather than the quantitative research of most social science. The FAUI project provides a unique opportunity to assess general hypotheses about government growth in systematic manner. With thousands of local governments in countries across the world, one can examine where and why they grow and decline, control multiple factors, and assess the distinct impact of each. These have been central themes in our FAUI conferences since 1983. Several chapters, but especially Chapter 5, present original U.S. results and draw on related work in other countries to illuminate government growth questions.

In our project meetings, discussions often start by observing differences across two countries—such as noting that France has a larger central government than the United States. The next question is why? One powerful idea is hierarchy, a general concept manifested by differences in class, income, and status. A central hypothesis to explain government growth is: the greater the hierarchy, the more the government growth. Rationale: government programs (of the left) seek to redress socioeconomic inequalities. They build programs to this end. As inequalities decline over time, however, so should demands for more government. We test this idea across U.S. cities in Chapter 2, with hierarchy measured by percentage black and inequality indexes of occupation, education, income, and national origin. Result: indeed cities with more hierarchy and disadvantaged residents have more active organized groups. And cities with more organized groups in turn have mayors and council members who are more fiscally liberal—they support increased spending. These fiscally liberal leaders in turn increased spending in the 1970s (Clark and Ferguson 1983), and as Miranda and Walzer report in Chapter 5, fiscally liberal council members increased spending in 1980-1984. But in some cities and time periods they analyze, this pattern does not hold. In part, it seems due to relatively greater national austerity and economic constraint in some years. Chapter 5 splices FAUI results with earlier Permanent Community Sample data to provide a 20-year perspective. One strong result is that as percent black residents increases, so does spending in subperiods of the 1970s, but this effect weakens in the 1980s. The 1970s results fit the general hierarchy proposition; the 1980s results lead us to consider countervailing mechanisms (e.g., of fiscally conservative leaders supported by taxpayer revolts).

Coping in Lean Years

U.S. cities were hit hard by grant cuts in the 1980s: federal grants to municipalities dropped by more than half after 1978. This brought dramatic change. Times are lean.

The Reagan administration sought to terminate many welfare programs, but Congress restored some and created a few new programs like JTPA (the Job Training and Partnership Act), replacing previous job training programs like CETA (Comprehensive Employment and Training Assistance). A major loss was general revenue sharing in the mid 1980s; it had provided no-strings-attached support for up to 10 percent of some city budgets. By contrast, more targeted programs are controversial. The Reagan administration extended the New Federalism approach of Richard Nixon: consolidate targeted programs into larger, general purpose block grants and delegate more autonomy to local officials for administration. Some

500+ specific grants were consolidated into block grants in the early 1980s. Many categorical programs were pursued unenthusiastically by cities, so loosening federal regulations, some argued, should raise grant "efficiency" in the sense of responding better to local citizen demand. Congress and national groups like the handicapped, for instance, had created categorical grants solely for handicapped programs for cities. Critics argued that decentralization would leave out the poor and disadvantaged.

The rhetoric of the Reagan policies, and its critics, often overshadowed actual changes. In his first year, 1980, President Reagan and top staff dramatically changed the Washington rhetoric from that of the Carter years that "nothing could be done." Federal income taxes were cut by one third in the early 1980s, and a New Federalism was proposed to eliminate many federal programs by transferring them to the states. But state and local officials refused most of these programmatic shifts and gained congressional support. Some Reagan staff, such as the OMB director David Stockman, targeted cuts in social programs in the early 1980s. By the mid- to late 1980s, Stockman was gone, as was most such rhetoric, and several programs began to grow. We have only a partial understanding of these changes as minimal urban policy research was completed in the 1980s and early 1990s. One Reagan casualty was the Office of Policy Development and Research of the Department of Housing and Urban Development, which in earlier decades supported many urban studies.

Clark and Walter in Chapter 6 survey specific changes in federal and state cuts and find several counterintuitive results. The Reagan rhetoric implied cutting the disadvantaged. But in fact federal grants in the aggregate became *more* targeted to low-income cities in the early 1980s, the opposite of the rhetoric and of evaluations of many individual programs. Presumably the evaluations of individual programs were correct, but Clark and Walter differ in using more comprehensive data. Nevertheless, the total dollar amounts from Washington were cut heavily, leaving the states with a larger role in making grants to cities. State aid is less redistributive (or "progressive") than federal aid. The combination of these two contradictory tendencies—more progressive federal targeting, but less direct aid, and a larger state role using less progressive criteria—produced only slight net changes in (combined federal and state) grant progressiveness during the Reagan years. Effects were also moderated by local political entrepreneurship, as some local officials (especially strong mayors and sophisticated staff) attracted more state grants. Although local officials often stressed development over redistribution, some dramatic cases of new programs focused on the most disadvantaged developed in the 1980s. The terms "homeless" and "underclass" entered our political vocabulary in these years. The disadvantaged became targets of new programs in cities where mayors made distinct commitments to the poor—such as Chicago under Mayor Harold Washington, Boston under Mayor Ray Flynn, and New York under Mayor David Dinkins. These were supported by local revenues, nonprofits, private foundations, and selected state and federal aid programs. Such initiatives were launched with a courageous tone as they so differed with the White House. But over time, more voices were added such that by 1992, Democratic President Bill Clinton was elected, who explicitly acknowledged his debt to big-city mayors like Dinkins and Daley.

Grant cuts were large in international context: the United States cut grants to cities in the 1980s more than any West European country we surveyed except Great Britain (see Mouritzen 1992, pp. 41-42). In aggregate, U.S. cities increased local revenues to make up for grant losses and kept spending increases just about equal to inflation. But both grant cuts and location of new revenues differed hugely across cities. Some states and cities raised taxes and fees substantially—a few more than replaced federal losses. Others did not. Fairbanks, Alaska, for instance, lost federal funds and saw big state cuts from a drop in oil revenues in the mid-1980s. Other depressed areas, such as oil towns in Texas, Oklahoma, Louisiana, and Alaska, saw home values drop by half and unemployment jump to 40 or 50 percent. In one short year, Fairbanks cut its staff nearly 40 percent in a dramatic battle between voters, unions, the council, and the taxpayer association. If Fairbanks-like drama is the exception, many cities still made substantial "adjustments," which we review in the next two chapters. The "simplest" adjustment is if grants are cut, the city cuts. Yet as Clark and Walter show in Chapter 6, few cities pursued this policy in such a mechanical manner; grant changes are less than deterministically related to local spending.

How can cities cope? Examples are throughout the book, but the last two chapters review specifics. Rowan Miranda's Chapter 7 offers powerful and specific results about contracting out. First, *contracting reduces expenditures and staff* for U.S. cities generally (he uses contracting data from the International City Management Association for

1,330 cities), although contracting increases average wages, perhaps because higher paid staff are kept to monitor contracted services and manage higher level work. Some questions remain about this first result, such as whether savings are passed on from the affected department and the direction of causality. But whatever qualifications limit this first result, its policy implications dramatically increase when combined with a second result: *contracting savings are primarily with nonprofits.* For 9 of 12 agencies studied, contracting with nonprofits brought savings. Contracting with other governments or with for-profit private firms brought fewer savings. Why? Nonprofits often operate with volunteer staff and foundation dollars, which reduce costs. "Privatization" with nonprofits may work better because city officials trust them more, especially if they are established organizations like hospitals or civic groups, for example. By contrast, city officials feel that new profit-making firms demand closer supervision. If nonprofits are less "profit oriented," they may be less likely to raise costs after an initial period of low bidding. Policy implication: if you are contracting mainly with for-profits or other governments, consider nonprofits.

Chapter 8 is on innovations that work. It reviews the 33 strategies from our FAUI survey. Some are widespread, such as *substantial movement toward user fees, less reliance on property taxes,* and *capital expenditure cuts.* Some cities have chosen *deferring maintenance,* but usually only in a limited manner—few cities have infrastructure problems like New York, with its infamous potholes and crumbling bridges. Money is tighter than a decade earlier, and local officials are far more sensitive to *productivity improvement strategies.* Of the many strategies, only some are innovative and only some work.

What innovations work? With local officials, participants in the FAUI project have identified innovative cities and leaders. One such effort is Urban Innovation in Illinois, which awards prizes to innovative cities, conducts case studies, and helps officials share innovations. Several creative innovations are reviewed, such as a joint agreement to form a tree-buying consortium; contracting out for computer acquisition, park construction, water meter inspection, and refuse removal; and a "shared savings" program to reduce energy costs.

Our FAUI surveys indicate international differences in strategies. What distinguishes U.S. cities? In the Reagan years did U.S. cities turn more toward strategies inspired by the private sector than cities in other countries? Only moderately, it turns out, for strategies like contracting out and user fees. Japanese cities rely twice as much on contracting out as U.S. cities. Although Swedish cities in the early 1980s scored just one eighth as high as Japan, Swedish scores increased dramatically by the late 1980s. Norwegian cities raised user fees more than those in the United States or any other country surveyed. All four Scandinavian countries launched "free city" experiments, loosening tight national regulations and encouraging local innovation. Despite their historic development of the welfare state, Scandinavian countries in the 1980s and 1990s made drastic grant cuts and major shifts in service delivery modes (contracting out, vouchers, market-based bidding, etc.). Considering the Scandinavian past, these were far more profound changes than those in the United States.

The FAUI strategy data permit us to answer questions of feasibility—where do different strategies work and why? Our survey results, and many case studies, indicate substantial entrepreneurship by local officials. The political culture of New Fiscal Populism increases sophisticated fiscal management practices and productivity improvement strategies. By contrast, innovation is usually unrelated to fiscal strain in U.S. cities. Put differently, the argument that local governments have too little money or legal autonomy to "cope" is often incorrect. Fiscal strain, income, and tax base measures usually show zero impact on innovation—that is, rich as well as poor cities innovate. This "liberating" finding should encourage less affluent cities not to feel condemned to past routines. Times are hard and lean for most cities. Austerity makes all of us suffer, but can sometimes spur innovation. Despite dramatic differences in history, culture, and legal arrangements, creative officials across the United States and in many countries are devising new ways to provide better services.

We had to jettison many past views in preparing this book; the reader may find some of his or her ideas challenged too. All chapters were prepared for this volume and are accessible to the general reader; technical issues are at chapter ends or in the Technical Appendix.

Note

1. The U.S. focus of this volume differentiates it from two more international volumes in progress that test similar hypotheses. The hypotheses about New Political

Culture are elaborated in a more extensive set of propositions (about 30 instead of the 6 here) and tested using the United States as one case along with about 10 others from Europe and Asia in Clark and Hoffmann-Martinot (forthcoming). A third volume differs from these first two in relying primarily on data from citizens rather than for cities; it is also international (Clark and Rempel forthcoming).

References

Clark, Terry Nichols. 1968. "Institutionalization of Innovations in Higher Education: Four Models." *Administrative Science Quarterly* 13,1 (June):1-25.

———. 1975. "Community Power." Pp. 271-96 in *Annual Review of Sociology*, Vol. 1, edited by Alex Inkeles. Palo Alto, CA: Annual Reviews.

——— and Lorna Crowley Ferguson. 1983. *City Money: Political Processes, Fiscal Strain, and Retrenchment*. New York: Columbia University Press.

——— and Vincent Hoffmann-Martinot, eds. Forthcoming. *The New Political Culture*. University of Chicago.

——— and Michael Rempel, eds. Forthcoming. *The Politics of Post-Industrial Societies*. University of Chicago.

Mouritzen, Poul-Erik, ed. 1992. *Managing Cities in Austerity*. London and Newbury Park, CA: Sage.

Quillian, Lincoln. 1990. "Political Parties, Interest Groups, and the SPO Thesis." Report by Fiscal Austerity and Urban Innovation Project, University of Chicago.

PART ONE

Old Cleavages and New Leaders

2

Race and Class Versus the New Political Culture

Terry Nichols Clark

Introduction

Scholars and political strategists have long debated class versus "nonclass" interpretations of politics and society. Karl Marx's scornful label "false consciousness" still haunts these debates. Should organizers mobilize on class lines? Yes, it advances the revolution, is the classic Marxist answer. Class is also most social scientists' answer to the research question "what are the primary sociopolitical cleavages?" This is explicit in textbooks and review articles on political behavior and social stratification (e.g., Lipset 1981; Coleman and Rainwater 1978; Vanneman and Weber 1987). Nevertheless, politicians and citizens increasingly disagree with this academic view of class. Social scientists, especially Americans, have been slow to change. Europeans have changed outlooks faster. Why? Perhaps because social changes in Europe are larger, most obviously in Eastern Europe and the former Soviet Union, but also in much of Western Europe, as we see below (and as elaborated in Clark and Hoffmann-Martinot forthcoming and Clark and Rempel forthcoming).

Social scientists debate "class versus nonclass" issues, often seeking universal answers and consistent commitment, with research and ideology substantially overlapping. A recent illustration is a vigorous exchange about the question "Are Social Classes Dying?" (Clark and Lipset 1991), in which ideological commitments sometimes led participants to reinterpret key concepts and speak past each other (part in Hout, Brooks, and Manza 1993; Pakulski 1993; and Clark, Lipset, and Rempel 1993). For Americans immigration and assimilation complicate class: the consensual U.S. view is that "minorities" can and should become "majorities." The Chicago urban sociologists of the interwar years (Park, Burgess, and their colleagues) saw these as obvious. Robert Dahl's *Who Governs?* (1961) and Herbert Gans's *The Urban Villagers* (1962) are just two social science examples of core models that stress assimilation. They focus mainly on white, European immigrants.

By contrast, for U.S. blacks or African Americans, assimilation models seem more problematic, as blacks have been in America longer yet enjoyed less success. Sociological theorists Orlando Patterson and William Julius Wilson found enthusiastic support among class theorists in publishing books entitled *Ethnic Chauvinism: The Reactionary Impulse* (Patterson 1977) and *The Declining Significance of Race* (Wilson 1978). Many interpreted these books as suggesting that race could and should be played down and class cleavages stressed for blacks. In the next decade, when Wilson's extensive empirical work documented the importance of race, a common reaction was that he had "changed his mind." This again illustrates popular ideological pressures toward a dichotomous commitment to either class or nonclass approaches.

Ideological debates are heightened in many national policy discussions as participants seek a single solution to serve as national policy for all cities—such as national health care or affirmative action or school vouchers. By contrast, an urban focus encourages intellectual pluralism; cities differ. Indeed, U.S. cities differ so much that if serious analysts want to make urban policy relevant to more than one locale, they must address city differences.

This chapter sets forth these points more powerfully than has been done in past work of which we are aware. To say that race causes conflict is too simple; in some countries, and some U.S. cities, this is false. Propositions here suggest where and why race sparks conflict, and our rich U.S. city data permit their precise testing. Results document the profound and enduring inequities of race and class cleavage as they shape the miseries or luxuries dividing urban Americans. We show how both class and race conflicts derive from hierarchy. The theory thus explains why race in the United States operates like class in Europe. International data dramatically illustrate this point (see Table 2.1). U.S. cities are analyzed with a variety of hierarchy/inequality indexes based on previously unpublished Census materials, which show a consistent pattern: race and class inequalities exert a profound and continuing impact on civic, neighborhood, and ethnic group activities; political leadership; and ideological outlooks. But comparative urban analysis also indicates the limits of these phenomena: in cities with few minorities and disadvantaged, a new political culture has emerged. Class politics has nearly vanished here. Or so it seems at first.

Rather than abandoning class for a nonclass "approach," we locate differences across cities with a theoretical framework that captures and explains them. We do not choose one side in debates over class, race, and nonclass views, or a "halfway" compromise, but show how different models are necessary to explain the workings of different cities. The chapter thus contributes to the sociology of knowledge as well as to urban policy analysis, by showing how different cultures define different legitimate rules of the game—for social science theories as well as political leaders and citizens.

Class Politics and the New Political Culture: Where Do We Find Them in U.S. Cities?

Our framework is stated in nine propositions that point to two patterns, two distinct processes: one toward the New Political Culture (NPC), a second toward "class politics," strongly colored by race. The NPC issues are environmentalism, growth management, feminism and abortion, gay rights, and other consumption and lifestyle concerns. By contrast, material and redistribution issues like pay, workplace conditions, and housing and health for the disadvantaged are more central to many black urban leaders.

One set of propositions suggests that the New Political Culture emerges more fully and forcefully in cities with less hierarchy (Proposition 2), more educated citizens (9), higher income (5), less poverty (3), and more high-tech service occupations (7).

A second process persists in other cities and neighborhoods: a U.S. variant of class politics, strengthened by the New Deal and Great Society programs and leaders. It is most powerful in cities with more hierarchical socioeconomic differences (Proposition 2), more nonwhite residents

(1) who encourage organized groups, and mayors and council members who articulate more redistributive policies (3, 4). Yet, these cities' resources are so constrained, by declining federal and state grants (see Chapter 6) and fiscally strained taxpayers, that their leaders' preferences are often not implemented in fiscal policies. The New Political Culture and class politics are often polar opposites: what encourages one discourages the other.

We find support for both sets of propositions. Although these trends characterize U.S. society in general, the two trends are mixed in national politics and by the media. The advantage of a local focus is to identify more pure and distinct examples of each process in specific cities, whereas the comparative urban data indicate how unique or general is each city. Still, if the two trends are conceptually distinct, the empirical findings are by no means deterministic. There is considerable choice and variation in how these issues are combined by citizens, group leaders, and elected officials. This "room for maneuver" is shown by moderate to weak statistical relations among key variables.

Class Politics: The Classic Left and Right

Most political systems have "progressive" and "conservative" tendencies. The terms left and right date from deputies' seating patterns under the French ancien régime. Left and right were not then linked to class politics; the lower classes generally could not vote. During the 19th century, class politics gradually identified left with the disadvantaged lower classes in Europe. This crystallized in the United States only in 1933 with Franklin Roosevelt's New Deal. We can distinguish its key elements with greater hindsight, now that they are eroding. Summarized in Figure 2.1, they stress conflict over workplace issues by upper versus lower class persons—in socioeconomic origins, economic activities, organized group activities, and political parties. Elements of class politics remain strong in some countries and U.S. cities, as we elaborate. But in many instances, class politics is being supplanted by a New Political Culture.

The New Political Culture, the Decline of Class Politics

A new political culture is emerging, but its recognition is long overdue, largely because many intelligent persons wear analytical

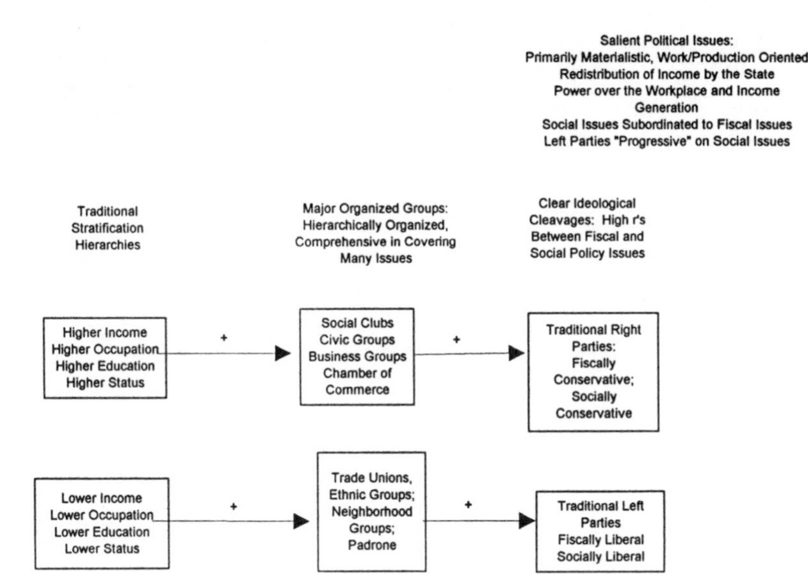

Figure 2.1. Class Politics: Major Components

lenses still focused on class politics. Class politics is not dead, but increasingly inadequate. Social stratification is expressed in new ways. Class never explained all the variance, yet was often the "best tool in the toolbox" and seemed adequate from the late 19th to mid-20th century in much of Europe and the United States. Since the 1970s, new nonclass cleavages have emerged in many countries concerning gender, race, regional loyalty, sexual preference, ecological concern, and citizen participation. These social issues are often neither fiscal nor economic; they may even cost government nothing. They concern new social patterns, cultural norms about how people should live. Most social issues began earlier, but their cumulative combination brings a fundamental change, a new political culture. Major components are in Figure 2.2, which contrasts with the class politics of Figure 2.1.

The New Political Culture (NPC) is distinct in seven key respects:

1. *The classic left-right dimension has been transformed.* People still speak of left and right, but definitions are changing. Left increasingly

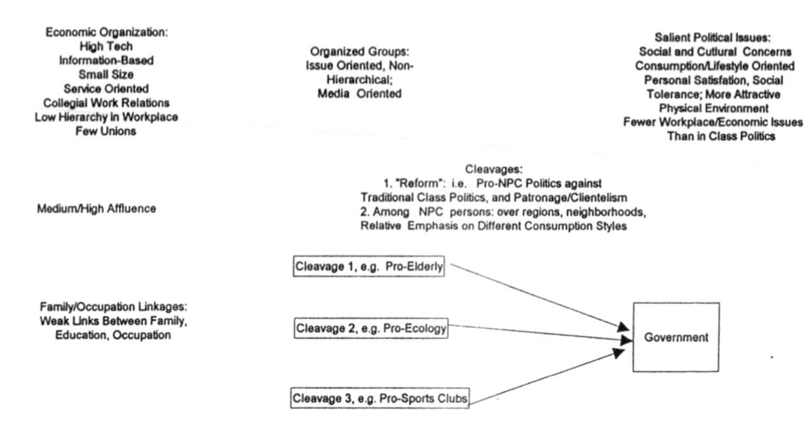

Figure 2.2. The New Political Culture: Main Components

means social issues, less often traditional class politics issues. In Eastern Europe the polarity of left and right so changed that by the late 1980s the political left sometimes referred to support for increasing private ownership and *less* state intervention in the economy. The change is less dramatic in the West, but increasing the role of government is no longer automatically equated with progress on the left, and the most intensely disputed issues no longer deal with ownership and control of the means of production. This transformation is under way in most countries of the world (compare Lipset 1991). Many leaders, and citizens even more, feel disoriented by shifting meanings of the left-right map.

2. *Social and fiscal/economic issues are explicitly distinguished.* Social issues demand analysis in their own terms. They are not just "ideological superstructure" or "false consciousness." Correspondingly, positions on social issues—of citizens, leaders, and parties—cannot be derived from their positions on fiscal issues. To say that they should be analytically distinguished does not imply that they do not overlap empirically. Social issues can have fiscal implications, such as providing extra funding for minority students. But social issues can also be pursued with no fiscal implications, such as Jimmy Carter's appointing more minority lawyers as federal judges. By contrast, the class politics model implies the opposite: (a) fiscal issues

dominate social issues, (b) social issue positions derive from fiscal issue positions, and specifically, (c) the left is liberal on social and fiscal issues and the right is conservative. We test these competing interpretations below, via correlations between fiscal and social issues. As class politics is replaced by the NPC, correlations between fiscal and social liberalism decline.

3. Social issues have risen in salience relative to fiscal/economic issues. Social issues ranging from ecology to women's issues to the right to burn the flag are increasingly evident. Their rise is driven by affluence: as wealth increases, people grow more concerned with lifestyle and other amenity issues in addition to classic economic concerns.

4. The rise of issue politics and broader citizen participation accompanies the decline of hierarchical political organizations. The NPC counters traditional bureaucracies, parties, and their leaders. "New social movements" and "issue politics" are essential additions to the political process. These movements encourage governments to respond more directly to interested constituents. By contrast, traditional hierarchical parties, government agencies, and unions are seen as antiquated. New demands are articulated by activist and intelligent citizens, who refuse treatment as docile "subjects" or "clients." They thus organize around new issues of welfare state service provision, like day care or recycling garbage. New groups seek to participate in general policy formation (rivaling parties and programs) and may press to participate in service delivery (rivaling government agencies, clientelist leaders, and unions). NPCs are thus seen as "rocking the boat"; they mean to. Conflicts with traditional particularistic leaders whose political support depends on clientelist patronage are particularly acute, for example, in Southern Italy or Chicago.

5. Market individualism and social individualism grow. Neither individualism implies a return to tradition; indeed the NPC clearly opposes the statist European right. Both individualisms foster skepticism toward traditional left policies, such as nationalization of industry and welfare state growth. But the NPC joins "market liberalism" (in the past narrowly identified with parties of the right) with "social progressiveness" (often identified with parties of the left). This new *combination of policy preferences* leads NPCs to support new programs and follow new rules of the game. Although NPCs vary, many support moderate state regulation (e.g., for ecological improvement), more so than classic market advocates like Milton Friedman.

6. *Questioning the welfare state.* Some NPC citizens, and leaders, conclude that "governing" in the sense of state central planning is unrealistic for many services, economic and social. Although not seeking to reduce services, NPCs question specifics of service delivery and seek to improve efficiency. They are skeptical of large central bureaucracies. They are willing to decentralize administration or contract with other governments or private firms—if these work better. "Work" includes citizen responsiveness as well as meeting professional staff criteria. In difficult economic times—like the 1970s stagflation—NPCs can become fiscally conservative, as specified in propositions below.

For many observers, right and left, lower level governments take on new meaning and can serve better than nation-states. We find struggles to develop smaller and more responsive governments and new intergovernmental agreements in the United States and many other countries. In East Europe, one of the first, dramatic reforms after the fall of communism in 1989 was reviving thousands of local governments in small towns that had been abolished in the past few decades. Many U.S. cities are criticizing their existing service delivery modes and experimenting with "alternative service delivery" patterns like contracting out, new technology, and so on. Neighborhood governments, block clubs, and the like are thriving the world over, simultaneous with declining turnout and interest in national elections (Nie, Verba, Schlozman, Brady, and Junn 1989).

7. *These NPC views are more pervasive among younger, more educated and affluent individuals and societies.* The NPC culture has emerged with basic changes in the economy and the family, and is both encouraged and diffused by less social and economic hierarchy, broader value consensus, and spread of the mass media, as we shall elaborate.

The Dynamics of Political Culture: Mapping Sources of Change

What drives the shift from class politics to the New Political Culture? Three general propositions are stated first, from which we derive several more specific ones. These are probabilistic ceteris paribus statements of relations, which identify factors causing changes in political culture, especially concerning the NPC and the decline of class politics. We thus begin one step prior to many political cultural analyses, by asking what determines political culture; what are the building blocks of different cultures—in particular, what factors are likely to vary across U.S. municipal governments in ways that will help explain differences in their relative support for different public policies?

A first principle does not directly address change in the culture of individuals, but identifies a major reason that nations and cities change: the people in them change. Persons who generally share preferences about a public policy are termed *sectors*. A demographically growing sector encourages policies consistent with that sector's preferences. The Demographic principle states:

1. *Demographic growth of a sector increases its legitimacy and power. The more persons there are in a political system (city or nation) who support a policy, the more likely it is to succeed.*

Analysts often use social background characteristics (like age cohorts) to identify sectors and possible corresponding changes in political culture. Demographic expansion of a sector can come from two sources, migration and differences in natality/mortality across sectors. Daniel Elazar (e.g., 1984) has developed the first approach, showing how distinct migration streams from Europe to the United States and across the United States helped implant distinct political cultures along the way. Second, change occurs if, for example, Catholics or fiscal conservatives have high birth and/or low death rates. Propositions below about blacks, poor, or more highly educated persons build on this Demographic principle.

2. *The more hierarchical the leading social institutions, the more likely powerful opposition parties are to emerge, challenging the hierarchies and pressing for egalitarian reforms. This is the Hierarchy Leveling principle: Down with hierarchies.*

This is a central proposition that we elaborate below. A related variable is the degree of legitimacy of the hierarchy. We can refine the Hierarchy Leveling principle by adding the more illegitimate the hierarchy, the more it sparks opposition. Legitimacy often seems weaker to those lower in a hierarchy; other factors weaken legitimacy too, although we leave them as exogenous for now. The Hierarchy Leveling principle operates most powerfully in a generally democratic context. Under an absolutist monarchy, dictatorship, or for example, in the antebellum South, it is hard for the Hierarchy Leveling principle to be expressed. There must be enough room for legitimate protest and political opposition to emerge. Otherwise opposition is limited to rumors, alcohol abuse, and dulled alienation by the many and anarchist or terrorist acts by the few.

Combining the Demographic and Hierarchy Leveling principles permits tracing political cultural reactions back to past and distant hierarchies that live on in the politics of their carriers. Immigrants from Russia, Eastern Europe, and the Middle East have helped spread terrorism and radical themes to Western Europe and the United States for generations. The closest U.S. analogue to the European feudal system was slavery in the South. As blacks migrated North, they brought their culture too. Although slavery was abolished in 1865, most blacks did not register to vote until after the civil rights victories of the 1960s. Political impacts followed—for blacks as well as others. Black politics in Northern cities shares many antihierarchical elements with European egalitarians: from Martin Luther King, Jr. to Harold Washington, a central issue was mobilizing the poor and disadvantaged through voter registration, mass demonstrations, and marches. Blacks thus added a radical tone to U.S. politics, elaborated in the 1970s by radical students, feminists, environmentalists, gays, and even the elderly via such labels as the "Grey Panthers" (Clark 1986).

These movements drastically changed the rules in many U.S. political institutions, from the U.S. Congress to cities and voluntary associations. Chicago is probably the most studied city in the world, and many past political analyses focused on an unusually hierarchical institution to explain most specifics: its Democratic party. Black mobilization against the party eventually redefined Chicago's political culture. But the situation evolved over several years, is still in flux, and is so graphic, that we chose to monitor it in an oral history with William Grimshaw over nearly a decade. Our interviews with activists brought out the hatred of former Mayor Daley and his Democratic machine. Tactical ploys to "unveil" its hierarchical, oppressive quality were consciously used again and again to mobilize black supporters. This came from the South: as an organizing tactic King typically chose openly racist and potentially violent places to organize, like Birmingham, Alabama, where Police Chief Bull Connor had his men use cattle prods against demonstrators. But as many blacks moved to Northern cities, the civil rights movement followed them.

To capture this cultural flavor, listen to Dorothy Tillman, Alderman and Democratic Party Committeeman in Harold Washington's Third Ward in Chicago in the 1980s and 1990s. She and another King coworker, Jesse Jackson, earlier helped mobilize black support against Mayors Daley and Byrne, and eventually persuaded Representative Harold Washington to run for mayor. How did they mobilize? Her answer: in the same way and with many of the same people as under King. Find a hierarchical opponent. Demonstrate against it. Try to force the hierarchy to use violence to stop your demonstration. Supporters will then rally to you.

> That's what they call me, the movement baby. I guess because I am from Montgomery and grew up with the Civil Rights struggle and I've been involved all my life.
>
> *How did you get up here?*
>
> The Headquarters called me to go on the northern tour, to go with Dr. King on the northern tour. So they picked their best. I was always like an advance person, you know, to go ahead. We toured the whole north to decide which city we were going to work in. . . . I didn't like it. But the consensus about coming to Chicago was that if we crack Chicago, we could crack the world. That it would all go across. (Tillman 1986, pp. 155-158).

King chose to organize in Chicago in 1966 in part because of the national image of Mayor Daley. It was most visible at the Democratic presidential convention held in Chicago in 1968, when Daley freely used police to break up street demonstrations. He, the national party, and Vietnam-scarred President Lyndon Johnson came under much political fire; they appeared in the national media as the major domestic resistance to peace and a more radical political program. But Daley learned from the bad press, and changed his style toward accommodation. As Tillman put it:

> Then Richard J. Daley said to us "You won't win here. The way you won in the South was they beat your heads. We're going to make sure

> they don't touch you here." And that's what he did. . . . I mean he was smart now. (Tillman 1986, pp. 155-158)

Daley moved symbolically away from hierarchy, as did many other governments and organizations in these years. This shift illustrates the *converse* of the Hierarchy Leveling principle:

2a. As hierarchy declines, it is harder to mobilize against it.

Flattening traditional hierarchies—of nobility and the estates, the church, the military, land and property ownership—has long been a program of the classic European left. Yet as these hierarchies themselves diminish in reality or in salience, their political importance for parties opposing them does too. Thus U.S. society with its more egalitarian traditions and past (absence of standing military, state church, aristocracy, etc.) encourages moderate proposals for social redistribution of income and status. To Americanize Hegel: no antithesis for want of a thesis. Opposition groups and parties generally weaken as hierarchy declines. Thus in countries, regions, or cities with more equal distributions of income, property, and status, one should find less conflict—at least hierarchical conflict.

Visibly and dramatically since the 1960s, organized group activities following the civil rights movement, anti-Vietnam war demonstrations, and unionization of municipal employees have brought more egalitarian styles to U.S. city governments. The same trend occurred after 1945, but often faster and more dramatically than in the United States, as France, Germany, and many countries in Western Europe and the rest of the world grew far more egalitarian than they were earlier, especially in styles of political leaders. They were more hierarchical until later, and caught up faster via international pressures to change.

But if the United States was more egalitarian in social relations and ideology than many European countries, its history of slavery joined to race has long made race a salient dimension of hierarchical social relations. Although slavery was abolished in the 1860s, blacks were denied the vote until the mid-1960s in much of the South and suffered many deprivations elsewhere. Overcoming hierarchial race relations is still current U.S. history. But rather than labeling U.S. "society" as racist, our propositions suggest looking for hierarchy and how it varies by subunits, for example, by region and by city. In those regions and cities marked by greater hierarchy in race relations, we should expect to find more race-oriented antihierarchical activities, such as civil rights groups and sit-ins to end discrimination. But in cities (and time periods) where racial concerns fall below a certain noticeable threshold, we should expect race-specific membership and activities to fall. This derives race-specific statements from our general hierarchy propositions.

Some of the most sophisticated recent research on race permits partial testing of these propositions, because they report results that vary by place and time (context). Studies of U.S. black voting and organized group participation generally show that mobilization is spurred by conflict and protest issues, more than for U.S. whites (Schuman, Steeh, and Bobo 1988). Conflict in turn we would expect to be triggered more by hierarchy and past racism. But this pattern shifts in U.S. cities with black mayors, whose presence dramatically signals that there is no longer a white political elite (see Boxes 2.1 and 2.2). This then shifts the rationale for race-specific organization and activities by many blacks. Consistent with Proposition 2A is the finding that black citizens in cities with black mayors exhibit participation patterns much closer to whites (Bobo and Gilliam 1990). Consider, however, one U.S. city with a black mayor, Los Angeles, which had the most extensive urban riots in the 20th century, when in 1992 a white jury ruled in favor of four white policemen who had beaten a black man. Lawrence Bobo and his colleagues conducted surveys tracking citizen attitude changes before and after this cataclysmic event. White attitudes toward blacks did not change materially. But blacks grew more hostile toward whites, for example, in response to the statement "American society has provided people of my ethnic group with a fair opportunity to get ahead in life." Those who disagreed rose from 49 to 63 percent; by contrast less than 10 percent of whites disagreed (Bobo, Johnson, Oliver, Sidanius, and Zubrinsky 1992, p. 8, Table C24). Presumably the events led to a heightened saliency of a race-specific hierarchy, which in turn triggered anger by blacks and feelings that the hierarchy was oppressive. But because the court decision had no such direct implication for whites, their more general attitudes about hierarchy did not change.

Another study examined the sources of organized group impact on city governments for 62 cities in the National Opinion Research Center's (NORC) Permanent Community Sample (PCS). It found that organized groups with large minority memberships had less

BOX 2.1 How Have Leading Black Mayors Articulated Class and Race Issues? Coleman Young, Mayor of Detroit

We are far from being the masters of our own destiny. We sure in hell don't control General Motors or Chrysler, Ford, or any of the major corporations that furnish jobs. We don't control insurance companies that charge us three times as much as they charge whites for automobile or home insurance. We don't control the businesses that have moved out of Detroit across Eight Mile Road and elsewhere in order to get away from blacks. Anybody who tells you something [different] is someone you ought to take a look at because that's a racist. Reminds me of something that Martin Luther King said, "How do you expect us to pull ourselves up by a bootstrap when we don't have boots?" The mothaf . . s stole our boots.

SOURCE: *Washington Post*, March 31, 1989, section A, page 3.

impact than other organized groups in 1975, controlling eight other group characteristics like membership size and level of activity. But when the same study was replicated in 1986, the minority membership effect fell to zero (Schumaker and Cigler 1989). This suggests that race grew less significant during this period. But more careful analysis of the same data suggest substantial variations by city (Hajnal and Clark unpublished).

Here and below we present results in terms of race, but our theoretical concern is to explain what specific factors associated with race in the U.S. urban context make it so important. The general answer is hierarchy, but because hierarchies change across cities and over time, so does the meaning of race. Race in the United States currently is negatively associated with most factors driving the New Political Culture; but if individual nonwhites change on the factors identified by the NPC propositions, they too should move toward the NPC.

Hierarchy can also vary by "situs," or separate hierarchical dimensions of stratification, such as occupation and education, as elaborated in Clark and Lipset (1991). The larger the number of hierarchical situses in a social unit (and the fewer the egalitarian situses), the greater the unit's overall hierarchy. A move toward less hierarchy in one situs encourages less hierarchy in others. For example, more collegial relations at work can spur demands for more egalitarian

BOX 2.2 How Have Leading Black Mayors Articulated Class and Race Issues? Harold Washington, Mayor of Chicago

Reagan and his mafia consider the cities a "special interest group." They believe we Democrats have "bought" our constituents with social programs. They're determined to undermine us among the poor, minorities, students, working people, by cutting off funds to these folks. That's what the big spending cuts is all about. It sure as hell isn't about budget balancing. And it isn't that self-help bullshit. It's politics, period. . . .

SOURCE: Miller (1989, p. 208).

politics and government. Or if racism persists in a firm or a male-only social club, more egalitarian government leaders may seek to change hierarchical policies concerning race or gender in these nongovernmental groups.

2b. Decline in hierarchy has in turn brought a decline in traditional left and right allegiances of citizens to political parties.

Although urban political parties have been receding since the late 19th century, they dropped further after the 1968 Chicago Democratic convention, when (still hierarchical) Mayor Daley used police against demonstrators (as discussed below under new social movements). We still test for effects of party labels of mayors and council members, but expect less impact than in a more hierarchical society such as France or in a (slightly) more hierarchical U.S. past.

3. The Consensus-Market Principle: The greater the value consensus among citizens, and the more that public policies reflect these values, the less citizens support organized groups, parties, strong political leaders, and a (hierarchical) bureaucratic state pressing for change.

Major differences among population subgroups are a classic reason for a strong state: to preserve order. This was Hobbes's Leviathan argument. But the polar opposite of the strong state is not inevitably his state of nature; markets, of various sorts, are also possible. Markets may depend in turn on modest state institutions—enforcement of contracts, rule of law, and so forth, as Spencer, Durkheim, and others stressed. These do not require a massive state hierarchy.

The Consensus-Market principle adds more power in combination with the Hierarchy Leveling principle (Proposition 2). A corollary follows: the legitimacy of hierarchy is maintained if many citizens feel that pressing problems would persist or grow worse if the hierarchy disappeared. This "need" for hierarchy is illustrated by the former Soviet Union and Yugoslavia, classic "nonconsensual" societies with multiple "republics, " religions, and language groups that long fought bitterly. The central state hierarchy persisted partly due to them. The Democratic party was unusually important in Chicago for analogous reasons: large numbers of people from the Austro-Hungarian empire migrated to this queen of the Midwest. When fundamental value differences decline—as they seem to have done in many Western European countries, and within many U.S. cities in recent decades—so should hierarchical states, parties, and related institutions. Less fundamental social and value conflicts imply less urgent support for a strong state to "right wrongs." This is consistent with Marx's observation that the state, as an instrument of repression, would disappear under socialism, or at least communism.

These general propositions identify factors leading from class politics toward the New Political Culture. The decline in hierarchies, stressed in Proposition 2, operates powerfully in all manner of institutions and societies. Critical in transforming U.S. politics are changes in the economy, the family, and education. Hierarchies have weakened in all of these.

4. Political issues shift with affluence: as wealth increases, people take the basics for granted; they grow more concerned with lifestyle and amenities. Younger, more educated, and more affluent persons in more affluent and less hierarchical societies should thus move furthest from traditional class politics.

This is a summary proposition; we examine its components below (for more detail, see Clark and Inglehart 1988).

Much evidence supports Proposition 4. Consider one result from our FAUI international survey that dramatically sets off the U.S. case. A good measure of traditional politics is a high, positive correlation between fiscal issues (should government spend more or not) and social issues (abortion, tolerance of minorities, etc.) That is, if you are fiscally conservative, this should determine your position on social issues, and place you on the traditional left-right continuum. Figure 2.3

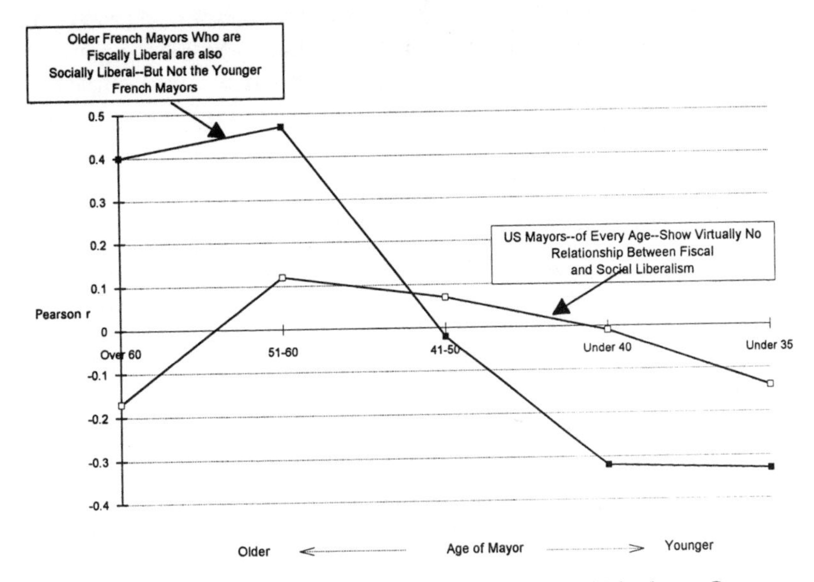

Figure 2.3. Only Older French Mayors Report Traditional Ideology: Correlations Between Fiscal and Social Liberalism for U.S. and French Mayors, By Age

shows that older French mayors (those over 50) show the traditional class-politics association between fiscal and social liberalism ($r = .4$). But for French mayors from age 41 to 50, the relationship drops to zero, whereas for those under 40, it flips to negative. By contrast for U.S. mayors of every age, fiscal and social liberalism are virtually unrelated in the 1980s. Pursuing this analysis in multiple regressions using nine variables (party, wealth, etc.), we find similar results: adjusted R^2s for the French are high (.25 to .44), about six times higher than for U.S. mayors (.03 to .08)! This survey dates from 1983; just a decade earlier, the U.S. mayors were close to the French (Clark and Ferguson 1983, p. 189). Hierarchy and class effects, these results suggest, have declined dramatically in a few short years.[1]

We pursue these ideas below, using hierarchy measures of income, occupation, race, and national origin. These measures, plus age, education, and ethnic group representation of both citizens and elected officials, are examined to see how they affect organized group activities and mayors' and council members' spending and social liberalism preferences.

Politics has long been analyzed as a consequence of class. We turn next to two areas—the economy and family—that once generated class relations, to identify a few key impacts on politics, "unpacking" ideas in Proposition 4.

Economic Organization Changes: Sources of a New Market Individualism

One simple, powerful change has affected the economy: growth. And economic growth undermines hierarchical class stratification.

5. Affluence weakens hierarchies and collectivism; but it heightens individualism.

With more income, the poor depend less on the rich. And all can indulge progressively more elaborate and varied tastes. But as this complexity increases, it grows harder to plan centrally; decentralized, demand-sensitive decision making becomes necessary. These contrasts particularly undermine centrally planned societies like the Soviet Union, as well as hierarchical firms like General Motors or U.S. Steel, and even IBM—all were in crisis in the 1990s. Thus:

6. Markets, ceteris paribus, grow in relevance as income rises.

Many private goods come increasingly from more differentiated and submarket-oriented small firms, especially in such service-intensive fields as "thoughtware," finance, and office activities. By contrast, huge firms are in relative decline, especially for traditional manufacturing products like steel and automobiles. Some two thirds of all new jobs are in firms with 20 or fewer employees in the United States and many other countries (Birch 1979). These small firms emerge because they outcompete larger firms. Why? Technology and management style are critical.

7. The more advanced the technology and knowledge base, the harder it is to plan in advance and control administratively, both within a large firm, and still more by central government planners.

Technological changes illustrate how new economic patterns are no longer an issue of public versus private sector control, but bring inevitable frustrations for hierarchical control by anyone. As research and development grow increasingly important for new products and technologies, they are harder to direct or define in advance for distant administrators of that firm, and even harder for outside regulators or political officials seeking to plan centrally (as in a Soviet 5-year plan, to use an extreme case). Certain plastics firms have as much as one third of staff developing the chemistry for new products. Computers, biological engineering, and robotics illustrate the dozens of areas that are only vaguely amenable to forecast and hence central control.

A major implication for social stratification of these economic changes is *decline in traditional authority, hierarchy, and class relations.* Current technologies require fewer unskilled workers performing routine tasks—or a large middle management to coordinate them—than did traditional manufacturing of steel, automobiles, and so on. High tech means more automation of routine tasks. It also demands more professionally autonomous decisions. More egalitarian, collegial decision making is thus increasingly seen as a hallmark of modern society, by analysts from Habermas and Parsons to Daniel Bell and Zbigniew Brzezinski, and to consultants in business schools who teach the importance of a new "corporate culture"—as illustrated by *In Search of Excellence* (Peters and Waterman 1982), the Number 1 nonfiction best-seller in the United States for some time, widely read by business leaders in the United States and Europe. *Fortune* magazine is filled with articles on similar themes; social scientists need to revise their class-based theories accordingly.

City measures analyzed below to capture these processes include per capita income, percent employment by sector (agriculture, manufacturing, and services), percent employed in professional/technical and related occupations, and percent employed in colleges and universities ("college towns" should be extreme cases of these patterns).

Education

The above economic changes expand occupations that are white collar, technical, professional, and service oriented. Occupational stratification increasingly resembles a diamond bulging at the middle rather than a pyramid. More education is needed for professional occupations; the numbers of students pursuing more advanced studies has rapidly increased in the United States since World War II and in Europe since the 1960s.

These increased educational levels have had huge political effects. They undermine the hierarchical, elitist nature of education that only a few decades ago distinguished university graduates as a small elite. More education also promotes more books, news, and media use, which in turn continue to spread new ideas after people stop their formal education. In recent years in the United States, the most widely read books have become works classified broadly as economics and sociology, supplanting novels.[2]

8. The higher the education of an individual, or average level of a social unit (country, city), the greater the support for classic issues of civil rights, civil liberty, and tolerance.

Education increases tolerance. Support thus grows with education for classic survey items like would you permit a Communist to speak to your school, or his or her book to be in the school library? would you accept a person of different social background as a marriage partner? Tolerance on such issues consistently increased from the 1940s to the 1990s in the United States and Europe (cf. Smith 1985; Weil 1985; Ward and Greeley unpublished; Davis 1989).

Education also has more impact when it exposes students to contrasting interpretations of what they know from their parents. Travel and more "cosmopolitan" social contacts have similar effects to education. Combined with travel and the worldly sophistication of television, these undermine the traditional intolerance of the parochial. When you understand the enemy, simplistic stereotypes of "foreign devils" become less persuasive. Although of course the media can also foster stereotyping, and in countries with controlled media, state and media leaders sometimes actively seek to manipulate public opinion.

What of the "New Right," and the success of militant groups, racial violence, and intolerance or the near success of former Klansman David Duke as candidate for governor of Louisiana in 1991? Such developments are undeniable. But the media understandably play up such incidents more than they do survey results of tolerance. These incidents are remarkable more for their relative absence in the United States, compared to the rise of anti-immigrant parties across Western Europe and heated ethnic conflicts in Eastern Europe and the Mideast. Some of these differences, in terms of the above propositions, seem related to the long immigrant history of the United States (demographic diversity) and relative affluence and high education level of the population. Nevertheless, in some U.S. cities and neighborhoods, one finds hostility and fear of gang warfare, drug battles, and crime that seem more like such third world countries. The Los Angeles riot of 1992 burst the bubble image of the United States as a Disney-like LaLa Land.[3] We explore the continuing importance of racial divisions in cities below.

Several longer term trends (growth of income and education) slowed with economic stagnation in the 1970s, and some reversed in the 1980s. Although our propositions posit simple linear relations, this does not deny that empirical social patterns sometimes cycle or reverse. These changes do not refute the propositions, only illustrate social complexity. And of course all the propositions are probabilistic; there are important exceptions and myriad other factors that also operate. The combined effects of these points decrease the observed coefficients.

An International Perspective on the United States

The U.S. results, to which we turn next, are highlighted by international comparison. Our FAUI surveys included a core of comparable items to several hundred mayors, currently pooled for seven countries. One question asked the mayor if he or she would prefer more, same, or less spending in several areas, from which we created an index of classic redistributive items—spending on public health and hospitals, low-income housing, and social welfare (see Chapter Appendix.) Then we analyzed characteristics of mayors and their cities to explain support for redistribution. The United States was strikingly different from many other countries. Note first that standard variables explain only modestly the policy positions of U.S. mayors—6 percent of the variance compared to 43 percent in Finland or 27 percent in France (Table 2.1). But the United States is not alone: Canada, Japan, and Australia are even lower than the United States in explained variance. Second, the powerful factors driving these policy preferences in France, Finland, and Norway are parties and class politics, measured by mayor's party and percent blue-collar residents in the city. By contrast *party is insignificant in the United States.* But third, documenting our propositions about race and hierarchy, *the most powerful factor explaining U.S. mayors' policy positions is the percent of black residents; mayors with more black residents favor more social welfare activities. Thus we may have less hierarchy and class politics in the United States compared to some European countries. But race is the U.S. version of a class-based party.*

TABLE 2.1 Why Do Mayors Support Social Welfare? International Contrasts

1. Three Countries Illustrate Traditional Class Politics: Party and Percent Blue-Collar Residents Drive Mayors' Preferences; Coefficients Are Strong and Multiple *R*s High

FRANCE

Multiple *R*	.61		
R Square	.37		
Adjusted *R* Square	.27		
Standard Error	.74		
N = 72			
Variable	*B*	*Beta*	*Sig*
Mayor's Party IPARTY3	–.01	–.44	.00
Woman Mayor IV146	1.07	.25	.02
Pct. Blue Collar IPCTBLUE	.03	.34	.03
Powerful Organized Groups	.01	.20	.07

FINLAND

Multiple *R*	.69		
R Square	.47		
Adjusted *R* Square	.43		
Standard Error	.54		
N = 167			
Variable	*B*	*Beta*	*Sig*
Mayor's Party IPARTY2	–.02	–.59	.00
Pct. Blue Collar	–.04	–.35	.00
Pct. Agriculture Empnt	–.02	–.19	.01

NORWAY

Multiple *R*	.35		
R Square	.12		
Adjusted *R* Square	.11		
Standard Error	.64		
N = 351			
Variable	*B*	*Beta*	*Sig*
Mayor's Party	–.01	–.29	.00
Population size	.08	.12	.04

continued

TABLE 2.1 Continued

2. Three Countries Illustrate Nonpartisanship on Welfare Issues: No Left-Right Party or Class Effects; Low or Insignificant B Coefficients and *R*s

CANADA

Multiple *R*	.38		
R Square	.14		
Adjusted *R* Square	.03		
Standard Error	.80		
N = 50			
Variable	*B*	*Beta*	*Sig*
Strong Party Organizations	−.02	−.38	.02

JAPAN

Multiple *R*	.33
R Square	.11
Adjusted *R* Square	−.03
Standard Error	.66
N = 73	
No variables significant	

AUSTRALIA

Multiple *R*	.34		
R Square	.12		
Adjusted *R* Square	−.01		
Standard Error	1.25		
N = 47			
Variable	*B*	*Beta*	*Sig*
Myr's Yrs in Office	−.09	−.31	.07

3. The United States Falls in the Middle: Race Performs Like Party and Class in France, Finland, and Norway, but With Less Powerful Effects

UNITED STATES

Multiple *R*	.38
R Square	.15
Adjusted *R* Square	.06
Standard Error	.89
N = 148	

continued

TABLE 2.1 Continued

Variable	*B*	*Beta*	*Sig*
Pct Black Population XBLK80US	1.07	.18	.05
Mayor's Years in Office MAYYRS	.04	.15	.08
Population Size (log) LIPOPT	.16	.15	.09
Strong Mayor (no Manager) Gvt. STRMAY	–.29	–.15	.10
Strong Party Organization SPOIX	.00	.10	.34
Pct. College Educated Res IPCTCOL	.01	.11	.35
Mayor's Party IPARTY2	–.01	–.07	.40
Mayor's Age IV144	–.01	–.07	.42
Pct. Agriculture Empl. IPCTPRI	–.02	–.07	.49
Powerful Organized Groups GRPSTR	.00	.03	.71
Per Capita Income LIPCINCT	–.15	–.04	.74
Pct Manufacturing Emp. IPCTSEC	.00	–.03	.77
Mayor's Education MAYED	.01	.01	.88
Woman Mayor IV146	.01	.00	.98
(Constant)	–.65		.87

SOURCE: FAUI surveys of mayors in each country supplemented by census and similar data.
NOTE: These are multiple regressions to explain support by mayors for increased spending on an index of social welfare, public health, and public housing, PRFVAR1. The Multiple *R*, *R* Square, and Adjusted R^2 are for full equations, but only independent variables significant below the .1 level are reported here to simplify presentation—except for the United States, where the full model is included. Similar model specifications were used in other countries. B = unstandardized regression coefficient; Beta = standardized regression coefficient; T = t-test; Sig. = significance (probability) level for that independent variable, where *significant at .10 level; ** at .05 level; *** at .01 level.

This international introduction spotlights race as we next elaborate in several interrelated processes for U.S. cities. Figure 2.4 shows the key processes. We consider specific steps in the Figure 2.4 model, moving from left to right, using the propositions above to assess how class politics and the New Political Culture vary across cities.

Testing the Propositions Using FAUI Data

We use the FAUI data files to measure key concepts from our propositions. The files include the FAUI survey (discussed in the

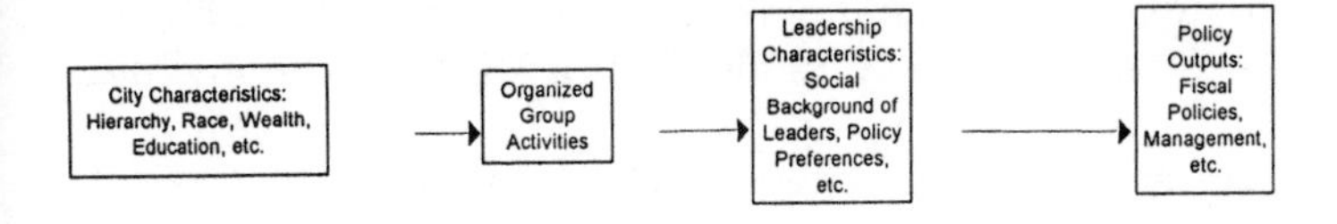

Figure 2.4. Framework of Key Variables

Series Introduction and Technical Appendix to the book) as well as other data, introduced below.

Socioeconomic Bases of Hierarchy

The importance of hierarchy in our propositions (especially 2, 2A, 2B) led us to invest considerable effort in developing original hierarchy measures for U.S. cities. We worked with several assistants merging data from 50 reels of mainframe tapes, one for each U.S. state, assembling highly detailed Census items for each city concerning the size of income groups (by $1, 000 categories), national origin groups (not only Italy and Poland, but even Crete and Gypsy!) and educational groups (number of men and women who had studied 11, 12, 13 years), and far more. We created city-level means, some simple ratios (such as rich divided by poor) for many variables, as well as more sophisticated indexes of hierarchy and heterogeneity, as suggested by Lieberson (1969), Atkinson (1975), Theil's entropy (1967), and others. See Coulter (1989) for an overview. More details on method are in the Chapter Appendix.

Income measures of hierarchy included:

- Proportion of the population below the federal poverty level (PERCPOV)
- The ratio of number of persons with income over $30,000 divided by the number with incomes under $10,000 (RATIOINC)
- The number of persons with income over $30,000 multiplied by the number with income under $10,000 (PRODINC)
- The proportion below poverty but not receiving public assistance income
- Four income versions of the Atkinson index of inequality

These income-based indexes of hierarchy all correlated highly with per capita income, but only moderately with percent nonwhites (Table 2.2).

The Atkinson index has been used in the past for income inequality. It follows the logic of a collective social welfare function based on degrees of inequality among subgroups; it permits assigning coefficients defining different criteria for assessing deviations from equality. The four versions we computed were based on varying such coefficients (see Chapter Appendix).

Education we analyzed in similar manner to income, using percent high school and college graduates by city as well as more detailed data on average years of education completed by city residents, from which we created four Atkinson inequality indexes for education.

National origin groups we classified by geographic regions (Northern, Southern, Eastern Europe, etc.) that sometimes correspond to broad cultural mappings—and to political coalitions in U.S. cities. We then used the percent of persons from each world region in each city to compute six versions of the Lieberson Index of Diversity, widely used by demographers to capture racial segregation—which can have obvious political implications. (Details in Chapter Appendix.)

Another way to analyze national origin is to rank countries of origin by status or prestige. We obtained prestige data from a national survey of U.S. citizens, the General Social Survey of the National Opinion Research Center (NORC), which asked "how desirable each of the following national groups is." We assigned these prestige scores to each national group in each city, and used them to construct Atkinson measures of inequality of national origin.

Finally, we created Atkinson indexes of occupational inequality, classifying occupations from high to low using another NORC survey of occupational prestige.

Clearly we had too many hierarchy indicators to continue through all analyses, but we generated several as we first used mainly race, which led critics to ask if this were an adequate test of the general hierarchy proposition (Clark, Balme, and Miranda 1988). There are strong interrelations among several hierarchy indicators, but moderately distinct dimensions. We do not advocate a view of stratification suggesting one basic dimension. Rather, consistent with the relativism of our political culture propositions, we note that citizens and leaders in NPC-type cities tend to focus on separate dimensions in defining political issues (e.g., women's issues, ecology issues, etc.).

Nevertheless, in cities where class politics is more salient, stratification is seen as more monolithic: hierarchies are seen as overlapping, with education, religion, and national origin all mutually reinforcing income. Leaders, as considered below, similarly vary along this unidimensional to multidimensional perspective.[4] The moderate correlations (for the entire United States) among these hierarchy measures support the interpretations of both distinctive and interrelated dimensions. Mean income, occupation, and education intercorrelate about .6; they correlate about .2 with the four Atkinson indexes; and race, on the bottom line, correlates .1 to .5 with all (Table 2.2). The Lieberson heterogeneity index was distinct enough from the Atkinson indexes to retain for subsequent analysis.

If the proportion of the minority population in a city, and racial inequality, were primarily identified with a single situs, such as occupation, then we should focus on that situs to explain racially oriented politics. Organization leaders like Martin Luther King, Jr., pursued this idea in targeting racist cities and government agencies for protest. In Chapter 4 we analyze specific patterns of "responsiveness" by government leaders to different groups, such as minorities, to analyze their effects (treating responsiveness as a measure of hierarchy in the government situs). But here we use primarily demographic and economic measures of hierarchy.

Interrelations among the several hierarchy/inequality measures are strong enough to indicate clearly that cities with more nonwhites have more inequality on income, education, and other indicators of hierarchy. But the relations are weak enough to refute two widespread views: (a) that "class" can be reduced to "race" or (b) race reduced to class. Each is a partially separate empirical dimension. We thus retain several separate hierarchy measures to analyze their separate impacts on the variables below. Our theoretical propositions about hierarchies are general enough to suggest patterns associated with each of these hierarchies, as we test next.

Does Hierarchy Increase Organized Group Activity for Minorities and in General?

How much do these hierarchy measures drive the political process variables? The first step in Figure 2.4 is from socioeconomic hierarchy to organized group activities. Do the hierarchy measures behave as expected? Yes, indeed. In cities with more hierarchy, the mayor and

TABLE 2.2 Correlations Among Hierarchy Indicators: Moderately Similar? Yes; Identical? No

	1	*2*	*3*	*4*	*5*	*6*	*7*	*8*
						MEAN-	*OCC-*	*LIP-*
	LIEBER4	*ATKINC2*	*GSSATK1*	*ATKTOT1*	*OCCATK1*	*EDUC*	*MEAN*	*CINCT*
1 Lieberson Index of Ethnic Diversity, LIEBER4								
2 Atkinson Index of Income Inequality, ATKINC2	.18**							
3 Atkinson Index of National Origin Inequality, GSSATK1	.39**	.25**						
4 Atkinson Index of Educational Inequality, ATKTOT1	.21**	.20**	.27**					
5 Atkinson Index of Occupational Inequality, OCCATK1	.01	.51**	.04	–.13**				
6 Mean Education of Pop (Yrs of Schl), MEANEDUC	–.24**	–.45**	–.18**	.06	–.61**			
7 Mean Occupation of Population, OCCMEAN	–.15**	–.42**	–.04	.28**	–.65**	.89**		
8 Per Capita Income (log), LIPCINCT	–.17**	–.61**	–.09**	–.07*	–.49**	.68**	.64**	
9 Percent Nonwhite population, NWH180	.50**	.31**	.16**	.48**	.11**	–.43**	–.30**	–.42**

SOURCE: US-FAUI survey supplemented by Census and other sources.
NOTE: These are *r*s, Pearson correlation coefficients, where * = significant at .05 level; ** = significant at .01 level. 1-tailed test. Atkinson and other indexes explained briefly in text and more technically in the Chapter 2 Appendix.

TABLE 2.3 Inequality and Race Generally Drive Low-Income and Minority Group Activities, Whereas Education, Occupation, and Income Increase Sierra Club Membership

	Group Activity Measures					
Hierarchy Indicators	*Activity of Low-Income Groups* V70R	*Activity of Minority Groups* V74R	*Average GroupSpend. Preferences* GRPPRF	*Average Group ActivityLevel* GRPACT	*Average Responsive-ness to All Groups* GRPRES	*Sierra Club Member-ship* AVGMEMSH
1 Lieberson Index of Ethnic Diversity, LIEBER4	.17**	.28**	–.05	.17**	.15**	–.01
2 Atkinson Index of Income Inequality, ATKINC2	.12**	.19**	.07	.08	.00	–.25**
3 Atkinson Index of National Origin Inequality, GSSATK1	.16**	.29**	.00	.16**	.12*	.04
4 Atkinson Index of Educational Inequality, ATKTOT1	.11*	.12*	.18**	.08	.04	.15**
5 Atkinson Index of Occupational Inequality, OCCATK1	.09*	.18**	.00	.07	.03	–.30**
6 Mean Education of Pop (Yrs of Schl), MEANEDUC	–.19**	–.27**	.00	–.14**	–.09*	.40**
7 Mean Occupation of Population, OCCMEAN	–.11*	–.19**	.05	–.07	–.05	.34**
8 Per Capita Income (log), LIPCINCT	–.14**	–.24**	.10*	–.10*	–.07	.43**
9 Percent Nonwhite Population, NWH180	.17**	.30**	.16**	.17**	.10*	.01

SOURCE: US-FAUI survey supplemented by Census and other sources for cities over 25,000.
NOTE: These are *r*s, Pearson's correlation coefficients, where * = significant at .01 level. 1-tailed test. Atkinson and other indexes explained briefly in text and more technically in Appendix.

council both reported that the city responded more to minority groups, low-income groups, and (less strongly) organized groups in general (see Table 2.3.) We analyzed these types of groups separately and in a summary index (GRPACT).

Moving from correlations to regressions, the same results hold. And the regressions indicate the importance of distinct independent variables, controlling for others in the model. Results: several hierarchy indicators increase activity levels of organized groups in general, of low-income groups, and of minority groups. Table 2.4 shows the regression explaining minority group activity: activity is higher in cities with high Atkinson inequality scores for education and national origin, more nonwhites, central cities (not suburbs), and less population growth. Some hierarchy variables intercorrelate enough to suppress one another when several are included in the same equation (e.g., in the Table 2.4 regression the Atkinson income index and college-educated persons remained below significance), but if entered a few at a time, they quite consistently increase group activity. As our main concern was to consider the general hierarchy idea in Proposition 2, these were useful alternative indicators of the general hierarchy concept. We thus did not seek to pursue the distinctive character of each separate "situs," that is, of income versus education and so forth.

Figure 2.5 shows first the national pattern (each FAUI city is a point in the scatter plot). Cities with more nonwhites (horizontal axis) are often higher on minority group activeness (vertical axis). Still the correlation of $r = .30$ reminds us that this is not a deterministic relationship. Some major cities are in the same figure, and more in Table 2.5.[5]

The hierarchy measures modestly increase activities of political parties in regressions using the same regression model, although effects are, as expected, weaker than for minority group activities (results not shown). Further, cities with powerful parties have weaker Sierra Club movements ($r = -.12$, significant at .01 level), illustrating the two alternative modes of political representation we have suggested.

Conclusion: hierarchy, measured in several ways, increases group activity for minorities and, in general, modestly increases political party activity.

New Social Movements: Environmental Groups and Related NPC Patterns

Some critics feel that "traditional" organized groups represent the establishment, in contrast to the "new social movements." Although one could argue that minority group activities really emerged as the first new social movement in the 1960s, they were in turn succeeded by women's groups, peace groups, environmental groups, and others

TABLE 2.4 Minority Organized Groups Are More Active in Cities With More Hierarchy

Dependent Variable: Activity of Minority Groups, V74R

Multiple *R*	.53			
R Square	.28			
Adjusted *R* Square	.25			
Independent Variables	*B*	*Beta*	*T*	*Sig*
Atkinson index of National Origin Inequality, GSSATK1	0.05	.25	4.64	.00 ***
Suburb (1 = Yes, 0 = No), SUBURB	–.073	–.26	–3.91	.00 ***
Percent Nonwhite Population, NWH180	2.18	.30	3.86	.00 ***
Atkinson Index of Occupational Inequality, OCCATK1	0.22	.16	2.05	.04 **
Atkinson Index of Educational Inequality, ATKTOT1	–0.07	–.14	–1.91	.06 *
Population Change 1980-86, POP8086	–3.34	–.09	–1.75	.08 *
Population Change 1970-80, POPCHG	–0.05	–.08	–1.72	.09 *
Traditional Political Culture, PKTRAD	–0.31	–.10	–1.71	.09 *
Proportion Age 35-44, AGE3544	8.67	.14	1.68	.09 *
Moralistic Political Culture, PKMORAL	0.21	.07	1.39	.17
Population Size (log), LIPOPT	0.12	.07	1.25	.21
Atkinson Index of Income Inequality, ATKINC2	–0.10	–.09	–1.06	.29
Per Capita Income (log), LIPCINCT	–0.50	–.08	–0.95	.34
Lieberson Index of Ethnic Diversity, LIEBER4	0.02	.06	0.92	.36
College-Educated Persons, V871	1.57	.11	0.79	.43
Proportion Age 25-34, AGE2534	–0.73	–.01	–0.26	.79
Mean Occupation of Population, OCCMEAN	–0.02	–.03	–0.19	.85
(Constant)	–3.14	12.89	–0.24	.81

SOURCE: US-FAUI surveys supplemented by Census and other sources for cities with a population greater than 25,000.

NOTE: Hierarchy is measured using several inequality measures shown here as independent variables, especially of Atkinson. Other variables from past urban research are included as controls. *N*(max.) = 1030. B = unstandardized regression coefficient; Beta = standardized regression coefficient; T = t-test; Sig. = significance (probability) level for that independent variable, where * = significant at .10 level; ** = at .05 level; *** = at.01 level.

TABLE 2.5 Cities Ranking High on Proportion Nonwhite Residents

City V796	Proportion Nonwhite NWHI80	Proportion Hispanic HIS80	Proportion Black BLK80	Minority GrpActivity V74R	Population 1980 POP80
COMPTON, CA	.98	.21	.75	.	84,685
EAST ST. LOUIS, IL	.97	.01	.96	.	53,381
LAREDO, TX	.93	.93	.00	.	99,874
EAST CLEVELAND, OH	.88	.01	.86	.	36,865
EAST ORANGE, NJ	.87	.03	.83	.	76,761
HIGHLAND PARK, MI	.85	.01	.84	2	25,733
BROWNSVILLE, TX	.84	.84	.00	2	91,440
HUNTINGTON PARK, CA	.83	.81	.00	.	48,495
MIAMI, FL	.82	.56	.25	4	382,726
MAYWOOD VILLAGE, IL	.82	.06	.75	.	27,766
LYNWOOD, CA	.81	.43	.35	.	50,809
INGLEWOOD, CA	.79	.19	.57	2	97,416
PICO RIVERA, CA	.79	.76	.00	.	54,800
GARY, IN	.78	.07	.71	.	147,537
NEWARK, NJ	.78	.19	.58	.	320,512
HIALEAH, FL	.76	.74	.02	0.5	154,713
PRICHARD, AL	.75	.01	.74	.	39,895
MONTEREY PARK, CA	.75	.39	.01	0.5	56,599
MONTEBELLO, CA	.74	.59	.01	.	54,594
WASHINGTON, DC	.74	.03	.70	.	633,425
CAMDEN, NJ	.73	.19	.53	.	83,942
EAST CHICAGO, IN	.72	.42	.30	4	38,458
MCALLEN, TX	.72	.72	.00	.	72,063
HARVEY, IL	.71	.05	.65	.	35,609
DEL RIO, TX	.71	.70	.01	.	31,855
HARLINGEN, TX	.70	.68	.01	.	47,083
CARSON, CA	.69	.23	.29	.	83,326
LA PUENTE, CA	.69	.62	.04	.	31,394
RIVIERA BEACH, FL	.69	.02	.67	4	28,437
ATLANTA, GA	.69	.01	.67	.	428,153
PLAINFIELD, NJ	.69	.07	.60	.	45,795
BELL GARDENS, CA	.68	.64	.00	.	35,827
GARDENA, CA	.68	.17	.23	4	46,070
UNION CITY, NJ	.68	.64	.02	.	56,931
EL MONTE, CA	.67	.61	.01	.	84,797
ROSEMEAD, CA	.67	.57	.00	.	44,213
WEST NEW YORK TOWN, NJ	.67	.63	.02	.	41,058
HEMPSTEAD VILLAGE, NY	.67	.09	.57	.	40,419
EL PASO, TX	.67	.63	.03	4	445,071
BELL, CA	.66	.63	.00	.	26,778
OAKLAND, CA	.66	.09	.47	4	344,652
DETROIT, MI	.66	.02	.63	.	1,138,717
BALDWIN PARK, CA	.65	.58	.01	.	53,835
ORANGE, NJ	.65	.06	.57		.
RICHMOND, CA	.64	.11	.48	.	75,927
PATERSON, NJ	.63	.29	.34	.	138,986
PETERSBURG, VA	.63	.01	.61	.	40,234
SOUTH GATE, CA	.62	.58	.02	.	72,015
SAN ANTONIO, TX	.62	.54	.07	4	819,021
NATIONAL CITY, CA	.61	.38	.09	2	53,687
GREENVILLE, MS	.61	.01	.60	.	41,022
NEW ORLEANS, LA	.60	.03	.55	3	564,561
DALY CITY, CA	.59	.19	.11	.	79,726
INKSTER, MI	.59	.02	.57	.	33,786
CHESTER, PA	.59	.02	.57	.	44,987
KINGSVILLE, TX	.59	.53	.04	4	29,870
OXNARD, CA	.58	.44	.06	4	115,657
SANTA FE, NM	.58	.55	.00	.	50,957
BIRMINGHAM, AL	.57	.01	.56	2	283,239
CHICAGO, IL	.57	.14	.40	4	2,997,155
BALTIMORE, MD	.57	.01	.55	.	774,113
PASSAIC, NJ	.57	.34	.20	.	53,342
SANTA ANA, CA	.56	.44	.04	0.5	217,219
WILMINGTON, DE	.56	.05	.51	3	69,896
ATLANTIC CITY, NJ	.56	.06	.50	.	37,857
HARTFORD, CT	.55	.20	.34	3	136,334
MOUNT VERNON, NY	.55	.05	.48	3	66,168

SOURCE: US-FAUI survey and Census.
NOTE: This shows just cities with more than 55 percent nonwhite residents. Number of cases = 67.

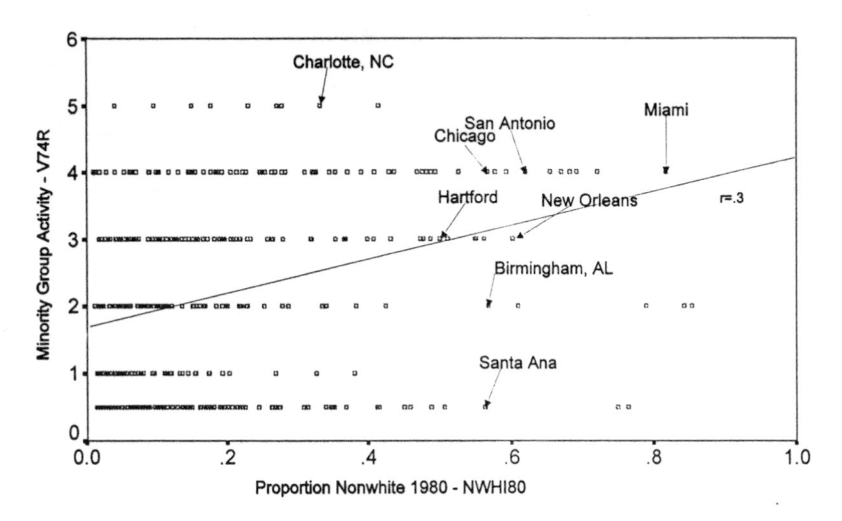

Figure 2.5. Cities With More Nonwhite Residents Have More Minority Group Activities
NOTE: See Table 2.5 for more cities.

in the 1970s. If minority groups often did not, these new movements consciously broke with the class politics of the day—specifically the leading groups on the left. Mayor Richard Daley represented the establishment Democratic party and excluded the new groups from entry to "his" 1968 Democratic convention in Chicago. Stung, the new groups started organizing immediately for the next presidential convention, and four years later elected George McGovern as their champion. Gloria Steinem was a leading spokesperson. The party program addressed the new issues. Daley, the party regulars, and unions lost out—and many voted Republican for the first time in their lives in 1972. This marked the end of the New Deal's dominance in national politics. Class voting in the 1972 presidential election plummeted to its all-time historic low for the United States on a standard class voting measure, the Alford Index, indicating that many blue-collar workers voted Republican, whereas many white-collar workers voted Democratic (Clark et al. 1993). They were led to switch by new issues and leaders, such as McGovern. Locally, similar dynamics were at work, with distinct movements and leaders often in open conflict with their predecessors (elaborated in Chapter 4).

How clearly can we detect new social movements locally? How much do they emerge following different dynamics from the other groups, as their proponents claimed? A decade after the new social movements were launched, the most consistently powerful was the environmental movement. Environmental issues were often the most salient issue in NPC cities in the 1980s and 1990s. But the activities were often distinctly local: neighbors would organize to clean up a park, volunteers would open centers to recycle newspapers, opponents to a new high rise would picket against it. But if these activities are often fueled by local individual initiative, some of the persons involved in them also work with national organizations pursuing similar ends. We thus contacted several national environmental groups, and from the Sierra Club acquired their membership and number of letter writers by locality. The Sierra Club conducted a survey of its members about their membership in other ecology and related groups in the 1980s. Members responding reported that on average they belonged to seven other ecological organized groups! Thus Sierra Club membership should indicate high ecology group activities of many sorts in a locality.[6]

We also sought data concerning other new social movements, such as activities of women's groups, gay and lesbian groups, and so on. Citizen data were easier to find than city-level data. We obtained only the *National Directory of Women Elected Officials, 1989* from the National Women's Political Caucus, which we added to our FAUI files. Results are below.

We have worked with students analyzing individual citizen data concerning membership in new social movements using several national survey sources, including the NORC-GSS for women's, gay and lesbian, and ecology groups, in Baldridge (1991) and Rempel (1992), plus attitudes concerning abortion, divorce, and family values in Butts (1992). Using these alternative methods to test the above NPC propositions, we find that many hold strong with citizen data, often stronger than the city-level results reported here. Results show quite consistently higher membership in new social movements, and support for NPC views, by persons who have more education and higher income and are younger, usually from larger urban areas, and sometimes Protestants or atheists—more so than Catholics or devout Protestants (especially for abortion and women's issues). These citizen survey analyses are being published as a separate book (Clark and Rempel forthcoming). Selected other measures for antigrowth and ecology movements are analyzed in Chapter 4.

TABLE 2.6 Sierra Club Membership is Higher in Cities With NPC Socio-economic Characteristics

Dependent variable: Sierra Club Membership, AVGMEMSH

Multiple *R*	.56			
R Square	.32			
Adjusted *R* Square	.31			
Independent Variables	*B*	*Beta*	*T*	*sig*
Population Change 1980-86, POP8086	5145.68	.13	4.18	.00 ***
Individualistic Pol Cul, PKINDL	359.51	.10	3.02	.00 ***
Moralistic Political Culture, PKMORAL	857.84	.29	8.40	.00 ***
Per Capita Income (log), LIPCINCT	2205.00	.33	6.27	.00 ***
Population Size (log), LIPOPT	115.11	.06	2.10	.04 **
College-Employed Persons/ Pop (log), LCOLLEGE	1803.80	.08	1.73	.08 *
Professional/Tech Employment/ 1,000 emp, PROFTCEM	2.30	.08	1.67	.09 *
Suburb (1 = Yes, 0 = No), SUBURB	98.24	.03	0.92	.36
Proportion Nonwhite Council Members, NWHICMBS	−348.64	−.02	−0.78	.44
Deep and Border South States, SOUTH3	−114.12	−.03	−0.68	.50
Population Change 1970-80, POPCHG	−11.12	−.02	−0.60	.55
Percent With Education 25 & Over w/4 yrs HS, PV870	−380.26	−.03	−0.53	.60
Traditional Political Culture, PKTRAD	−25.29	−.01	−0.16	.88
Constant	−26136.62	3219.46	−8.12	.00

SOURCE: US-FAUI surveys supplemented by Census and other sources for cities over 25,000.
NOTE: *N*(max.) = 1030. B = unstandardized regression coefficient; Beta = standardized regression coefficient; T = t-test; Sig = significance (probability) level for that independent variable, where * = significance at .10 level; ** = at .05 level; *** = at .01 level.

Sierra Club membership, consistent with Proposition 4, is higher in cities with more professional and technical occupations, colleges, higher per capita income, and more population growth 1980-86 (Table 2.6 regression). These cities were more likely to have political cultures characterized by Daniel Elazar as moralistic (including persons from the New England migration stream through Minnesota to the Pacific Northwest) and individualistic (Elazar's Middle Atlantic culture). Below significance were college-educated persons, cities in the South, and cities with Elazar's traditional political cultures. Most findings thus fit the NPC propositions quite nicely.

Conclusion: cities with younger, more educated, affluent, white residents in professional and technical occupations have more group activities like those of the Sierra Club, which champion consumption and lifestyle issues for their residents.

Leadership Patterns: Race Drives Most Leadership Differences

The next step in our Figure 2.4 path model is from organized groups to leadership, measured by social characteristics and policy preferences of elected officials. First, social characteristics, and to pursue the propositions, we consider leaders' racial characteristics in association with the redistributive policy preferences of class politics. Then, pursuing NPC leaders, we look at those who are younger and more highly educated to see if they report more NPC policy preferences (Proposition 4).

Which cities elect more black mayors and council members? Black mayors, our regressions show, are more likely in cities with more black citizens, more income hierarchy, less population growth, more Sierra club membership (showing that ecology and black leadership *can* go together, especially in liberal locations like Seattle, New Haven, or Berkeley where a large white majority elected black mayors). Results from Models 1ABCD are in the Chapter Appendix. Here and below we mainly summarize regression results rather than displaying the dozens of tables; the Chapter Appendix defines models such as 1ABCD.

Black council members, unlike mayors, do *not* increase in number with the proportion of nonwhite citizens (in regressions). But they do appear more in larger cities with more ethnic segregation. Our segregation measure (the Lieberson Diversity Index) specifically captures the concentration of blacks, Hispanics, and distinct national origin groups that thus seem to advantage minority council, more than mayoral, candidates. Most other variables fall below significance in regressions (specified using the same Models 1ABCD as for black mayors).

Concerning other social characteristics, more educated and women mayors are more socially liberal (measuring social liberalism by support for abortion, sex education in schools, and gun control, SOCCONS). Younger mayors were also more socially liberal in correlations, but not regressions.

We analyzed two measures of fiscal policy preferences:

- General fiscal liberalism, indicated by support for spending on 13 policy issues, FISCONS
- Welfare spending preferences, summing support for spending on welfare, hospitals, and public housing (the index used in international comparisons in Table 2.1, PRFVAR1)

Race was important for both: nonwhite mayors supported more spending on both measures. Cities with more organized group activities and more income hierarchy also had mayors who supported more welfare spending, showing the impact of earlier variables in our model. But in general, we were impressed at how few nonracial characteristics were significant in explaining fiscal policy preferences, although we tested more than 20 (see Model 2 in the Chapter Appendix).

Figure 2.6 shows the strong and consistent differences in mayors' spending preferences between cities with high versus low percentages of nonwhite citizens. Figure 2.7 shows the same pattern for organized groups: cities with more nonwhites have more organized group activities of all sorts (even taxpayer associations), with one exception: Republican party activities. Similarly, individual citizens are less active and influential in cities with more nonwhites.[7] These patterns of policy preference and organized group activity clearly illustrate the greater importance of organized groups in cities with left and ethnic political culture, consistent with the formulation in Clark and Ferguson (1983) (elaborated below). Clark (forthcoming) presents case studies of mayors in four cities with more ethnographic detail of their leadership patterns.

What about political parties? We noted low Republican activeness in cities with high nonwhites in Figure 2.7. Republican mayors are modestly more fiscally conservative ($r = .12$). But in regressions with other variables controlled, membership in more fiscally conservative parties has virtually no impact on mayors' fiscal or social conservatism. This U.S. pattern differs sharply from other countries like France or Finland as we saw in Table 2.1. Only two decades or so ago, however, the pattern was quite different in the United States as well.

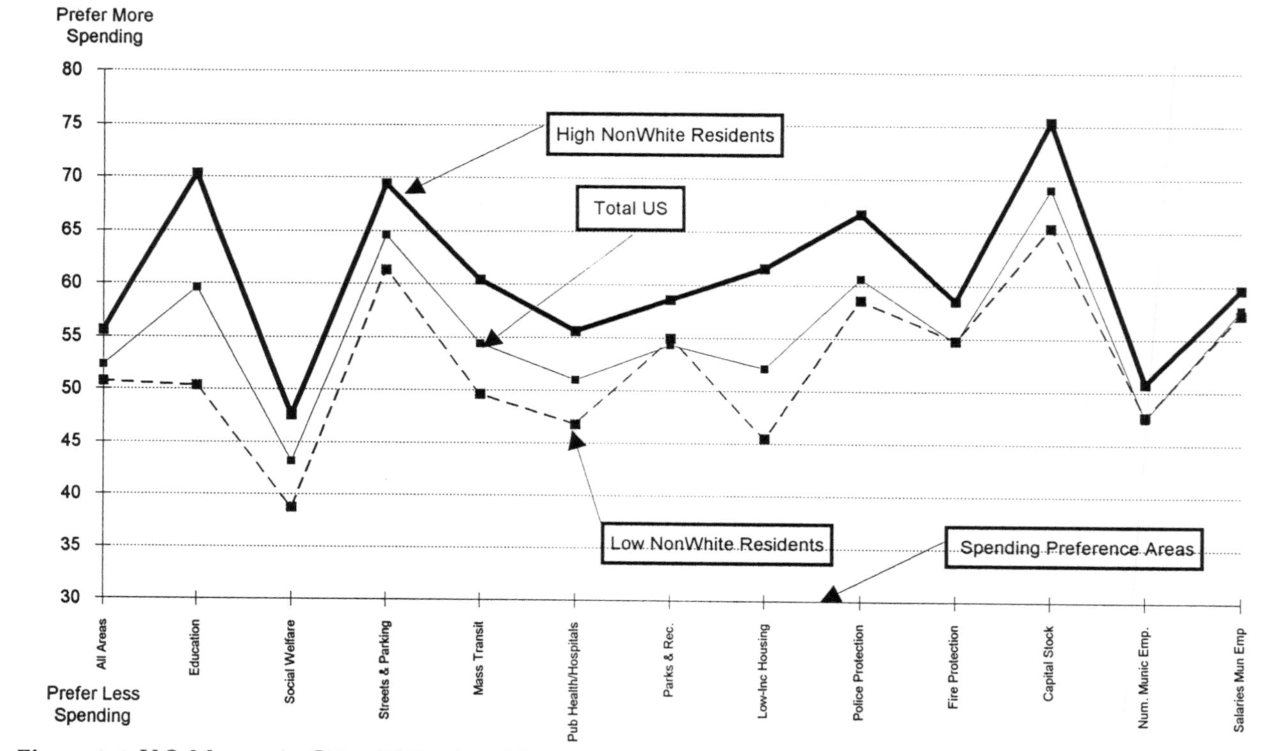

Figure 2.6. U.S. Mayors in Cities With More Nonwhite Residents Consistently Report Higher Spending Preferences
See Figure 2.7 note

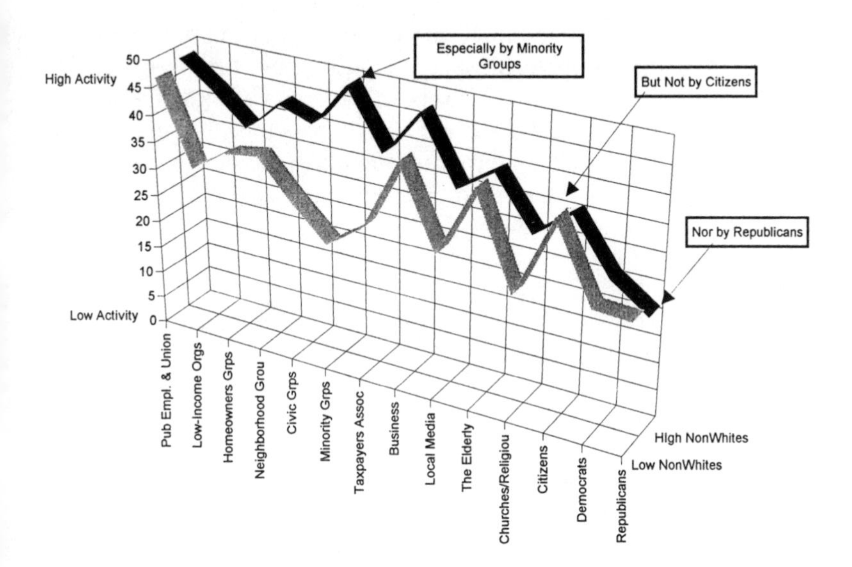

Figure 2.7. Mayors in Cities With More Nonwhite Residents Report More Activity by Most Organized Groups

SOURCE: US-FAUI survey supplemented by Census and other sources for cities over 25,000.
NOTE: Cities were divided into those with more than 30 percent nonwhite residents and others with less than 5 percent nonwhite residents.

Regression coefficients for effects of parties on fiscal and social conservatism were then close to the European pattern (Clark and Ferguson 1983, p. 189; Clark and Inglehart 1988). This indicates the rise of NPC patterns in the United States during the past 20 years.

Our propositions suggest a less clear "left-right continuum" backed by party discipline for NPC mayors. Instead we expect NPC mayors to break from this New Deal continuum and respond more to specific issues. To capture issue specificity we asked mayors about "their own preferences" for spending first in "all areas of city government" and next in 12 specific spending areas (Figure 2.6 lists these). Mayors responded for all 13 items on a scale from "spend a lot less," to "spend a lot more." PRFSD is the standard deviation computed for each individual mayor across the 13 specific spending items. The rationale: the traditional left supports more spending on all areas of government, and the traditional right supports less spending. PRFSD operationalizes the first definition of the New Political Culture listed above: "the traditional left-right dimension has been transformed." How so? Consider a mayor who consistently prefers more spending on all 13 items, and a second who consistently prefers less spending. The standard deviation (PRFSD) of both across the 13 items is thus zero. But a third mayor, who prefers spending more on some items but less on others, will have a higher standard deviation across his or her items. A mayor high on PRFSD, then, is more NPC oriented because he or she breaks from the consistent spending preferences of both traditional left and right.

For each mayor we computed a PRFSD measure of "issue specificity." Our general hypotheses about the New Political Culture suggest which mayors are more issue specific in preference patterns; we tested these in regressions explaining PRFSD. Three things were significant. Mayors reported more issue specificity if they had held office for fewer years and were in towns where more residents had college degrees and the industrial sector was larger. The first two findings are consistent with Proposition 4. The third is less clear in meaning. In simple correlations, further variables were important: mayors serving fewer terms in office or fewer years as mayor, per capita income, and a citizen importance index all increased issue specificity of the mayor. But these were too similar and collinear with other variables to hold up in regressions.[8]

Given the multidimensional nature of leadership, it is useful to characterize cities by how they combine several dimensions, which together constitute the "rules of the game" that define political culture. Cities differ in political culture. We distinguished four types in Clark and Ferguson (1983): left/Democratic, right/Republican, Ethnic, and New Fiscal Populist. We created scores of each of the four political cultures for each city by summing five more basic dimensions or deep structures: fiscal liberalism, social liberalism, organized group activities, populist responsiveness to citizens, and concern for public versus separable goods. Each of these five was measured by surveys of each mayor and council member; the five were then summed to generate a score for each city. The "transformation rules," or algorithms, for combining the five into the four political cultures are in Table 2.7. (Note that left/Democratic and right/Republican political culture are here measured solely by these five deep structures, not by party affiliation as analyzed above.) New Fiscal Populism is defined as fiscally conservative, socially liberal, less responsive to organized groups, more responsive to citizens, and more

TABLE 2.7 General Characteristic (Deep Structures) Generating Four Types of Political Culture

	Policy Preferences		*Legitimate Sources of Input to the Political System*		*Public Goods Emphasized as Resources*
	Fiscal Liberalism	*Social Liberalism*	*Individual Citizens*	*Organized Groups*	
Traditional Left	+	+	–	+	–
Traditional Right	–	–	+	–	+
Clientelist/Ethnic Politics	+	–	–	+	–
New Fiscal Populism	–	+	+	–	+

NOTE: These four types were developed in Clark and Ferguson (1983). The New Fiscal Populist (NFP) differs from the New Political Culture primarily in the NFP's greater emphasis on fiscal conservatism.

favorable toward public goods. What cities have such mayors? The simple, clear results: New Fiscal Populism is stronger in cities with more white citizens and council members, and white mayors. Most other variables fall below significance in regressions specified with the same model used for spending preferences (Models 2ABCDE).

What of ethnic political culture? It is defined as the near converse of New Fiscal Populism, measured by fiscal liberalism, social conservatism, more group activities, less populist response to citizens, and use of clientelist-separable goods rather than public goods (Table 2.7). Cities score higher on ethnic political culture if they have more nonwhite citizens, a nonwhite mayor, more nonwhite council members, more active organized groups—especially minority groups—and more hierarchical income distributions (ATKINC2) (again in regression Models 2ABCDE).

Conclusion: these results document starkly how much race divides cities in leadership patterns and policy preferences, as it did previously for organized activity patterns. It is impressive how little impact other variables had—the variables discussed in much past work on urban politics—such as reform government, declining population, and several others tested in our regressions.

What of Policy Outputs?

We consider three classes of policy outputs: fiscal policies, management policies, and development strategies. Do the above differences contrasting the New Political Culture and nonwhite class politics recur in policy outputs? Perhaps.

The mass media regularly point to striking differences in selected cases. For instance East St. Louis, Illinois (population 55,000, 96 percent black in 1980), suffered dramatic urban ills for much of the 1980s. The financial condition was so bad that the city hall building was sold, and more than one mayor and many council members left office in recurring scandals. The major issues were classic poverty-related concerns: jobs, poor housing, crime, drugs, welfare dependency, and so on. Many cities with high minority populations tend in this direction.

At the other extreme stand cases like Princeton, New Jersey, where the most visible public issue in the early 1990s was how to cope with an expanding deer population. Deer were causing traffic and other problems. Using hunters was clearly unthinkable. Finding a "politically correct" policy took multiple public debates; many themes were introduced flowing from New Political Culture concerns. The final solution was birth control "pills," administered in the deer's food. Conspicuous by their absence were *all* the issues salient in East St. Louis.

The stark differences separating East St. Louis from Princeton recur in many cities. Variations across localities have been a classic theme in the United States since Tocqueville. These were documented through the 1970s by several studies showing systemic patterning of differences in policy outputs, often explained by local decision making and sociodemographic variables (e.g., Clark and Ferguson 1983). Repeating some of these analyses for the 1980s, however, we were surprised at the magnitude of the changes, using several standard measures of fiscal policy outputs—revenue, expenditure, and debt. Differences across cities in the 1980s thus grew distinctly less patterned than a decade or two earlier. There was far less research on cities in the 1980s and early 1990s than previously, so we do not have a flow of studies on specifics as earlier. But analyzing variables used in earlier years, it is surprising how few had any significant impact on policy outputs, with one exception: grants from state and federal governments. Grants became the major variable driving local fiscal policy. This marks a major change from earlier years. It is true of 16 fiscal policy measures: total revenues, general revenues, own revenues, and debt, for 1980 to 1985, 1985 to 1989, 1980 to 1989, and levels per capita in 1989. Again and again, we analyzed many of the same potential explanatory variables that we and others used in the past, as discussed in the leadership and organized group analyses just reported. The only consistent results: (a) grants had a major impact on level and changes in spending, and (b) little else was consistently significant (using regression Model 3 in the Chapter Appendix).

One might reason that grants should be important as they have been cut back so drastically: federal per capita grants to cities dropped by more than half from 1978 to 1991. Conversely, one might reason that as grants constitute a smaller portion of local revenue, they should explain local fiscal policy *less*, and other characteristics should explain more. This may be true for some specific issues, but not for the big fiscal policies we analyzed.

An alternative interpretation: the more important factor driving decisions may not be grants and their cutbacks per se, but a more general "marketization" of decision making. That is, city governments increasingly act and compete in an environment defined less by federal and state governments and their grants. In the 1960s and 1970s, total local spending was enhanced by grants for specific activities, some poverty oriented. Grantsmanship skills of local officials were important, and developed substantially in these years. But during the 1980s, local policy makers shifted from grantsmanship to focus more on individual citizen/voters, featuring productivity and avoiding taxpayer revolts. Leaders must demonstrate skills like managing scarce resources, contracting out when it is more cost-effective, and generally responding less to the traditional Democratic, fiscally liberal political culture. Shifts from grantsmanship to austerity-driven management are largest in cities that traditionally had Democratic and ethnic local leaders, whereas those with Republican and New Fiscal Populist leaders have changed less; the others just moved in their direction. If Harold Washington, for instance, as the first black mayor of Chicago, had been elected in 1968, he would very likely have pursued and received large federal grants for the disadvantaged. But serving in the most austere Reagan years (as mayor from 1983 to 1987) he faced what his planning commissioner Elizabeth Hollander characterized as "a national political climate [of] New Fiscal Populism" (1986, p. 178).[9] He and his staff were correspondingly austere in their fiscal policies; especially in his early years he was substantially more austere then his immediate predecessor Mayor Jane Byrne, whose spending increases nearly generated a fiscal crisis. The Harold Washington example illustrates how national constraints trimmed many fiscal sails.

One policy rewarded in this climate is "sound management." If money is scarce, managing limited funds carefully stretches each dollar. Candidates campaign on irresponsibility and excesses of their opponents, and propose more sound management; New Fiscal Populists have been particularly successful in this regard. We included several items in our FAUI survey of the chief administrative officer, asking about the degree to which the city government used revenue forecasting, performance indicators, and the like. We summed these in our Fiscal Management Index (FMIX) and examined its variation across cities. General results fit with NPC leadership propositions. Cities with more sophisticated fiscal management have residents who are more educated and higher in income; the cities are larger and have more professional city managers, staffs that are more influential than the council, and councils that share fiscal policy preferences with their mayor (in regressions using Models 2ABCDE, adding selected variables).

We also analyzed several economic development activities, as surveyed by the International City Management Association (ICMA) in 1984 and 1989 (and other related surveys discussed in Chapter 4). "Traditional" economic development incentives—such as tax abatements, loans, rezoning, land write-downs—were only weakly or not at all related to most socioeconomic and leadership characteristics. The forces of "marketization" seem to have encouraged such competition among municipalities (and states) that most have adopted broadly similar packages—at least enough so that differences remain hard to explain, except by more idiosyncratic characteristics like an unusually creative mayor or local plant closing. One development program does, however, fit the more general pattern of this chapter: cities with more nonwhite (and especially black) citizens and black mayors and council members were, not surprisingly, more likely to support minority development programs (item defined in Chapter Appendix). This held strong in regressions.

Other policy outputs that vary across cities are the FAUI 33 strategies. These have been analyzed, for example, as subsets of Expenditure strategies, Revenue-Raising strategies, and Productivity strategies, and all 33 combined. Productivity strategies were measured using items like "improving productivity through better management" and "improving productivity by adopting labor-saving techniques." Cities, we found, adopted more productivity strategies if they had more sophisticated fiscal management (on the FMIX measure), performed more functions, had mayors who were more powerful and New Fiscal Populists, had fewer council members who were Democrats and more who perceived "problems" chosen from a list of 11 (like declining tax base, unemployment, and pressure from local taxpayers). The strategies are discussed further in Chapter 8 of this volume.

Conclusion

This chapter outlines propositions specifying factors shifting the rules of the game by which cities are governed, concentrating on changes from class politics toward the New Political Culture. The propositions stress the general concept of hierarchy as sparking the dynamics of both race and class conflict. Hierarchy, as relativized in these propositions, thus transcends specific arguments about race "versus" class. The propositions identify central elements driving each toward political mobilization. Propositions are tested with data from the Fiscal Austerity and Urban Innovation Project. Several original measures of hierarchy are analyzed, which indicate that separate hierarchy dimensions remain empirically distinct (with *r*s below .5). Indeed, race and class are just two of many hierarchical dimensions of social stratification. Although separate dimensions overlap somewhat, many have clearly separate and distinct effects on sociopolitical processes.

Consistent with the propositions, the New Political Culture is more common and powerful in cities with less hierarchy, more educated citizens, higher income, less poverty, and more high-tech service occupations. By contrast, a U.S. version of class politics is more common in cities with more hierarchical socioeconomic differences, more nonwhite residents, more active organized groups; their mayors and council members articulate more redistributive policies. These results fit the propositions quite well. Hierarchies spark political organizations, conflict, and ideological differences among leaders.

However, as we shift from leaders and their policy preferences to actual policy outputs, we find far less clear support for these distinctive patterns. Differences across cities recede, at least compared to past years. Cities are so constrained by declining federal and state grants, and by fiscally strained taxpayers, that leaders' preferences are often not implemented in fiscal policies. We do find distinct differences across cities in management policies, such as computerization of budgeting, more common in NPC-type cities, as expected. But many traditional economic development policies differ little across cities—because the policies have become so widely used. Two development policies do differ, however, in ways that match our propositions. Minority development programs are more common in cities with minority residents and leaders, and various ecology-sensitive and growth management policies are more common in NPC cities, as elaborated in Chapter 4.

Appendix

The Atkinson Inequality Indexes

Social science has a long tradition of inequality indexes, including the Lorenz curve, Gini coefficient, Theil's entropy, and more. They help operationalize our concept of hierarchy. Atkinson (1975) contributed by joining social welfare theory with statistical measurement to generate a flexible but powerful methodology for capturing the degree of hierarchy in an income distribution. Like other social welfare economists, he used income as a simple quantitative aspect of social welfare. We in turn extended the approach to related social hierarchies, such as occupation and education. Atkinson's index increases with inequality, but how much depends on the coefficient alpha. The researcher can set the alpha in the defining equation:

$$I_r = 1 - \left[\sum (y_i/\overline{Y})^{1-\alpha}(P_i) \right]^{1/(1-\alpha)}$$

where

- I_r = Atkinson's Index of Inequality
- Y_i = income of income class i (e.g., persons from $5,000 to $10,000 annual income)
- Y = mean income for the social unit (the city)
- P = proportion of income earned by income class i
- α = alpha = the exponential coefficient which the analyst can vary to specify rate at which inequality affects the overall Index

If one wishes to emphasize, or deemphasize, the effect of a small improvement in income by the poorest income group, one can vary the alpha coefficient accordingly. Although Atkinson and most interpreters discuss this in terms of "the research's choice," in our terms one might assign alpha values to match different political cultures—a left political culture is presumably more sensitive to the lowest income groups than conservative political cultures. Accordingly we chose four alpha values in computing each of our Atkinson indexes: .5, 0, –.5, and –1, a range of alpha values recommended by Atkinson and researchers who have refined the index (e.g., Bartels and Nijcamp 1976; Schwartz and Winship 1979).

We created 16 Atkinson indexes, with four alpha scores each for income, occupation, education, and national origin. However, results for our U.S. cities showed that indexes using each of the four alpha score were quite similar (most *r*s over .9), so we retained only one index for income, occupation, education, and prestige of national origin, four in all. Thus despite the theoretical appeal of assigning different alpha coefficients to the Atkinson index, these U.S. city data suggest that such (slight) variations in method for measuring inequality/hierarchy generate minimally different results.

The four Atkinson indexes for income, occupation, education, and prestige of national origin were in turn moderately interrelated (see Table 2.2), but not enough to generate substantial multicollinearity; all were below .5 except occupation with income ($r = .51$). Further computational specifics:

The 17 *income* categories for annual per capita income in 1980 came from the STF4A U.S. Census tapes which ranged from:

01 Less than $2,500 to

.

.

17 $75,000 or more

The income (and education) data were available for men and women separately, but at the city level these were too similar to analyze separately for men and women.

The 14 *education* categories were for years of schooling completed by city residents age 25 and older in 1980 from STF4A and ranged from:

01 0 No schooling

02 1 to 4 years of elementary school

03 5 to 7 years of elementary school to

.

.

14 6 years or more of college

The 14 education categories were also available by sex but we again found them too similar to analyze separately for men and women.

For *occupation*, we obtained data for each of our approximately 1,000 cities for the proportion of employed persons working in a large number of occupations from the STF3 Census file. To rank the occupations we used the occupational prestige scores from the National Opinion Research Center's General Social Survey (NORC-GSS) summarized in "Computing 1989 Occupational Prestige Scores" by Keiko Nakao and Judith Trease, provided by Tom Smith at GSS/NORC. Here are the occupational categories and prestige scores:

Variable	*Score*	*Occupational Category*
XT66_1US	53.52	Executive, Admin., and Managerial Occupations
XT66_2US	64.38	Professional Specialty
XT66_3US	51.21	Technicians and Related Support Occupations
XT66_4US	35.77	Sales Occupations
XT66_5US	38.16	Admin. Support Occupations, Including Clerical
XT66_6US	27.84	Private Household Occupations
XT66_7US	48.40	Protective Service Occupations
XT66_8US	30.93	Other Service Occupations
XT66_9US	35.57	Farming, Forest, and Fishing Occupations
XT66_AUS	38.51	Precision Prod., Craft, and Repair Occupation
XT66_BUS	33.36	Machine Operators, Assemblers, and Inspectors
XT66_CUS	35.94	Transportation and Material Moving Occupations
XT66_DUS	29.44	Handlers, Equip. Cleaners, Helpers, and Laborers

The Atkinson Occupation Index these generated was OCCATK1.

For *national origin* we again used the relative status of country of origin from another NORC-GSS survey that asked how desirable different countries were. They ranged from English 6.46 and French 6.07 to Mexican 3.52, in D. Huffer programming files GSSATK.PAD and INDEXES3.MEM.

The Lieberson Indexes

The leading indexes of diversity of social groups were developed by Stanley Lieberson (e.g., 1969). He and others often used the measures to compare neighborhood segregation. But the indexes similarly permit identifying the degree to which a city is concentrated in its ethnic composition, for instance to contrast a city with many different ethnic groups all equal in size with a second dominated by a single ethnic group. The basic Index of Diversity, A_b, formula is

$$A_b = 1 - \sum [P_i Q_i]$$

where

P_i = the proportion in category i of population P
Q_i = the proportion in category i of population Q

and P and Q are two different populations for which one may use any number of i categories, from 2 to infinity. For P we used the U.S. national average for all of our cities, calculating as i the proportion of each national origin group. Then each city was treated as Q, and the degree to which it differed from the national average in its ethnic concentration was represented in its Diversity score. We used five algorithms, varying the number of ethnic groups from the 49 available in the STF4A Census file (LIEBER5D), down to six major ethnic groupings important for city politics purposes (black, Hispanic, Asian, white, Western European, and Eastern European), LIEBER4. Most analyses used LIEBER4 because using a larger number of national groups diminishes the relative importance of blacks and Hispanics—politically inappropriate for the years we are studying, in which they are often the most important ethnic groups.

FAUI Survey Items

The FAUI survey of all U.S. cities with more than 25,000 population included mayors, council members, and chief administrative officers. See the Series Introduction and Technical Appendix for more information. The three full questionnaires are published in Clarke (1989, pp. 253ff). Question numbers (e.g., Mayor Q9) refer to the questionnaire, whereas Variable numbers (e.g., V70R) refer to the *Fiscal Austerity and Urban Innovation Project*

Codebook prepared by Robert Stein (1985). Bickford, Clark, and Ritchie (1989) is an introduction to the survey and index construction.

Mayor and Council Spending Preferences

The key FAUI spending item was Mayor Q4: "Please indicate your own preferences about spending. Circle one of the six answers for each of the 13 policy areas. 1 Spend a lot less on services provided by the city 2 Spend somewhat less 3 Spend the same as is now spent 4 Spend somewhat more 5 Spend a lot more DK Don't know/not applicable. Policy areas: All areas of city government, Primary and secondary education, Social welfare, Streets and parking, Mass transit, Public health and hospitals, Parks and recreation, Low-income housing, Police protection, Fire protection, Capital stock (e.g., roads, sewers, etc.), Number of municipal employees, Salaries of municipal employees."

FISCONS = fiscal conservatism, sum of scores by the mayor for spending on each of the 13 areas

PRFVAR1 = factor score sum of spending preferences for the mayor on social welfare, public health and hospitals, and low-income housing

Group Activity and Influence Indexes

Mayors and council members were given a series of three items about different participants: their spending preferences, activity, and council responsiveness to each group.

Mayor Q7: "Please indicate your judgment about the *spending preferences* of several participants in city government affairs. Circle one of the six answers for each of the types of participants. Does the participant want to: 1 Spend a lot less on services provided by the city 2 Spend somewhat less 3 Spend the same as is now spent 4 Spend somewhat more 5 Spend a lot more DK Don't know/not applicable." This was followed by 21 participants, including neighborhood groups, and others listed in Figure 4.4 of Chapter 4.

GRPPRF was the average spending preferences by the 11 civic and other groups (all participants shown at the base of Figure 2.7 of this chapter except citizens and the two parties)

PARTYPRF was the spending preference score for the two political parties

MYCV7D was the absolute difference between the mayor's overall spending preference and that of the council = ABS (V7R-V290R).

Mayor Q8: "Please indicate *how active* the participant has been in pursuing this spending preference. Circle one of the six answers for each of the types of participants. Has the participant carried on 1 No activity 2 Little activity 3 Some activity 4 A bit of activity 5 A lot of activity 6 The most activity of any participant DK Don't know/not applicable." The respondent then scored 21 participants. Participants left blank were coded at the bottom of the activity level.

Activity variables analyzed include three items in Models 1ABCD: Average Group Activity Level GRPACT (computed for the same 11 groups as GRPPRF above), Activity of Low Income Persons V70R, and Activity of Minority Groups V74R.

Mayor Q9: "Please indicate how often the *city government responded favorably* to the spending preferences of the participant in the last three years. Circle one of the six answers for each of the types of participants. The city has responded favorably 1 Almost never 2 Less than half the time 3 About half the time 4 More than half the time 5 Almost all the time DK Don't know/Not applicable." The same 21 participants were listed as in Q7 and Q8. Participants left blank were recoded at the bottom of the scale.

Responsiveness Indexes for the same 11 groups as GRPPRF above: Average Responsiveness to all groups as reported by council members, GRPRESC and by the Mayor, GRPRES.

SOCCONS summed three social liberalism/conservatism items: Mayor Q16 (V128): Would you favor or oppose a law which would require a person to obtain a police permit before he or she could buy a gun? (Circle one number) 1 Favor 2 Oppose 3 Don't Know.

Q19 (V131) Would you be for or against sex education in public schools? (Circle one number) 1 For 2 Against 3 Don't Know/Not Applicable.

Q20 (V132) Do you think abortion should be legal under any circumstance, legal only under certain circumstances, or never legal under any circumstances? (Circle one number) 1 Under any circumstances 2 Under certain circumstances 3 Never legal 4 Don't know/Not Applicable.

FMIX = Fiscal Management Index = sum of five CAO FAUI items (Q9) concerning sophistication of revenue forecasting, fiscal information systems, performance indicators, accounting and financial reporting, and economic development analysis. The full items are published in Clarke (1989).

Four Political Cultures

The political culture variables were constructed from mayoral responses to individual survey items which were joined in five indexes. The logic for combining the dimensions is shown in Table 2.7. The social conservatism index (SOCCONS) uses responses to questions about handguns, sex education, and abortion. The "populism" index uses questions about the responsiveness of government to citizens and how often leaders take positions unpopular with their constituents. An index of interest group activity sums items from Mayor Q9 about how often the government responded favorably to public employees, low-income groups, homeowner groups, neighborhood groups, civic groups, minority groups, taxpayer groups, elderly groups, businesses, and church and religious groups. Public goods orientation is measured by response to a question about whether or not leaders should break some rules to help people. Finally, fiscal conservatism (FISCONS) comes from mayors' reports of their own spending and tax preferences. Exact index construction is reported in Bickford, Clark, and Ritchie (1987, pp. 62-65). Every city for which there is adequate survey data is assigned a score for each of the four political cultures. NPCIX is the New Fiscal Populism Index.

Regression Models

Recursive regression models were used to explain each block of variables shown in Figure 2.4, except that each block was estimated including most other causally prior variables. Variables were drawn from past work by ourselves and others such as Clark and Ferguson (1983), the Fiscal Austerity and Urban Innovation Project annual *Research in Urban Policy,* and Clarke (1989). Variables not identified above are briefly described here. Most regressions were estimated with pairwise deletion of missing cases, which retains the full number of cases available for each bivariate relationship. With as many as 1,030 cases, this meant that for the Census variables we often had relatively complete *N*s, whereas analyses including variables from the FAUI and ICMA surveys had lower *N*s, as their response rates were 40-65 percent. Practically all important results we checked to see how robust they remained in several ways. Occasionally we would try listwise deletion for this purpose. Practically always we would compare results with simple correlations and look for consistency with the regression coefficients. Usually we compared results across varied regression specifications. For instance, by reestimating simpler regression models, deleting subsets of variables that were somewhat multicollinear and which had low *N*s to see if key findings consistently held. We do not seek to summarize all variations estimated, but list below the core regressions generating results reported in the text (which the text often identifies by an acronym such as "Models 1ABCD"). The Technical Appendix at the end of the book considers sample selection bias.

Models 1ABCD all included

1. Atkinson Index of National Origin Inequality, GSSATK1
2. Percent Nonwhite Population, NWHI80
3. Atkinson Index of Educational Inequality, ATKTOT1
4. Population Change 1980-86, POP8086
5. Population Change 1970-80, POPCHG
6. Traditional Political Culture, PKTRAD (Elazar's 1984 type, coded by city; see Kincaid 1982)
7. Moralistic Political Culture, PKMORAL (Elazar's type)
8. Population Size (log), LIPOPT
9. Atkinson Index of Income Inequality, ATKINC2
10. Lieberson Index of Ethnic Diversity, LIEBER4
11. Proportion Age 25-34, AGE2534
12. Mean Occupation of Population, OCCMEAN (from same data used in Atkinson occupation index above)
13. Sierra Club Membership, AVGMEM (from Sierra Club)
14. Average Group Spending Preferences, GRPPRF
15. Proportion of Single and Divorced Persons in City, SNGLDIV

Then because five organized group activity measures were too highly interrelated to include together in a single model, they were added one at a time to each of four separate models:

Model 1A Average Group Activity Level GRPACT
Model 1B Activity of Low Income Persons V70R
Model 1C Activity of Minority Groups V74R
Model 1D Average Responsiveness to all groups as reported by council members, GRPRESC and by the Mayor GRPRES

Model 2 added several leadership-related variables to test for their impact, estimated in five ways (Models 2A, 2B, 2C, 2D, 2E in NPC68BD.PAD). 2A used 20 variables that led to moderately biased results with so many included together; to look for consistency, 2B and 2C each used just half of the 2A variables, as follows:

Model 2B:

1. SPOIXStrong party organization index (see Miranda this volume)
2. PARTYPRFMean party spend preference
3. FPIX84NCFunctional performance index 1984 = range of functions performed by the city (see Clark and Ferguson 1983 and Technical Appendix below.)
4. LPARTYDIlog of party ideol dist (using Mayor Q7 for Democrats minus Republicans)
5. INDEPMayor is Independent (not Demo or Rep)
6. LDEMCMB2Proportion Democrats on council
7. MYCV142MPopulism Index, mayor/council avg(V142R+V437R)/2
8. LPROBALLIndex of 11 urban problems from CAO
9. FOGForm of government (mayor vs. manager)
10. MISPARTYNonresponse to party id question
11. PARTYRESMean govt responsiveness to party

Model 2C:

1. PARTYPRFMean party spend preference
2. WOMANMAYWoman Mayor from Natl Dir Woman El. Offls. 1989
3. HISPCMBSProportion of Hispanics council
4. WMCMBSProportion women couclmbs=v470
5. NWIMAYNonWhiteMyr, v145, (Wh, old2=0)(1,
6. MYAGEMayor's Age=V144w(99=sysmis)
7. MAYYRSNumber of years as Mayor
8. MYEDMysEducat, SteinCode=V148, 99=sys
9. MYSEXMMysSex, 1 = M, 2 = Fm(N = 47), 9 = sysmis from FAUI survey
10. NWHICMBSProp nonwhite counclmbs = (V460, 2, 3, 4).

Models 2D and 2E both added to 2B and 2C

1. AVGMEMSHSierra Club AvgMembership
2. NWHI80Pop nonwhite 1980
3. GRPACTAverage group activity index
4. ATKINC2Atkinson index of income inequality

Model 3 was specified to help explain policy outputs, including as dependent variables 16 fiscal measures: total revenues, general revenues, own revenues, and debt, for 1980 to 1985, 1985 to 1989, 1980 to 1989, and levels per capita in 1989. Other policy outputs included International City Management Association survey results for 1984 and 1989, FAUI strategies, and others. Independent variables:

1. MYCV142M Populism Index, mayor/councilavg(V142R+V437R)/2
2. NWIMAY NonWhiteMyr, v145, (Wh, old2 = 0)(1, 3, 4, 5 = 1
3. MAYDIX1 Importance of Mayor, Open-Ended (V111 ADDTO V115/5) See Chapter 4 Appendix for item.
4. NWHICMBS Prop nonwhite counclmbsmbs = (V460R+V462+3
5. FPIX84NC Functional performance index 1984
6. MYED Mayor's Educat, SteinCode = V148, 99 = sysmis, 7/91
7. MAYYRS Number of years as Mayor
8. SOCCONS MY-Social Conservatism Index (0 +V128R +V131R +V132R)/3
9. SPOIX Strong party organization index
10. LDEMCMB2 Proportion Dem council members = V418S/(V417+8 See Chapter 4 Appendix for item.
11. NPCIX MY-New Fiscal Populist Pol. Cult. Index
12. COVDIX1 Import of city council, Open-Ended(V111 ADDTO V115/5) See Clark/Goetz, Appendix 1, for item.
13. MEDIA MEDIA USE: (IV140+IV141)/2

    Source: FAUI Items: Q28 (IV140) "How often did you use the local media (radio, TV, the press) in the last two months of your last campaign? Please include both paid advertisements and other media coverage.) (Circle one number) 1 Name appeared in the media several times daily 2 Name appeared about once a day 3 Name appeared about once a week 4 Name appeared less than once a week 5 Don't know/Not applicable." Q29 (IV141) "Excluding election periods, how often have you appeared in the local media in the past two years? (Circle one number)" [Same five response categories as Q28].
14. FOG Form of government (mayor vs. manager)
15. EPCIX MY-Ethnic Political Culture Index (see text of chapter)

ICMA Minority Development

Several ICMA Economic development items are summarized in the Chapter 4 Appendix. The Minority Development item was Q11: "Does your city have any special programs to encourage economic development projects undertake by minority groups? YES___ NO___ If YES, please indicate those programs that are part of this effort: (check all applicable) a. Technical assistance to minority business b. Financial assistance programs established specifically to help minority business c. Commitment to a specified level of city procurement from minority business d. Other (specify)_____."

Similar items were posed about zoning, land use, tax abatements, loans, and the like. Source: International City Management Association, Economic Development-1984.

Notes

1. The full results are in Clark and Inglehart (1988).

2. In the United States, book reading has nearly doubled during the past 40 years. Asked "Do you happen to be reading any books or novels at the present time?" in 1949, 21 percent said yes; this jumped to 37 percent in 1990. Further, the number of new books, titles, and editions saw nearly a fourfold increase in "sociology, economics" from 1959 to 1989 (4 to 15 percent), replacing "fiction" as the first-place category, which fell from 16 to 11 percent in these years. Surveys by the Gallup Organization, and data from U.S. Census and *Publishers Weekly*, compiled in *The American Enterprise* (1991, p. 92).

3. Some studies also suggest a decline in social liberalism among young persons in the late 1980s and early 1990s. See Clark and Rempel forthcoming.

4. Social stratification studies report similar differences in perception of social stratification: higher status persons perceive that society is composed of a multiplicity of distinct status dimensions, whereas lower status persons perceive stratification as essentially based on income alone (e.g., Coleman and Rainwater 1978).

5. Blacks, Hispanics, and Asians share hierarchical backgrounds, so we analyze them summed using the Census "nonwhite" category. Blacks and Hispanics comprise the great majority of nonwhites in U.S. cities, but overlap little: $r = -.10$. A larger proportion of each increases minority group activity, although blacks ($r = .29$) more than Hispanics ($r = .12$), which with *N*s of 415 are still both highly significant. For proportion nonwhite $r = .30$. Similar patterns recur in multiple regressions controlling for income, education, and other variables.

6. Sierra Club membership and letter writers correlated over .9, so we used simply membership. The Goetz antigrowth movement, analyzed in Chapter 4, and Sierra Club membership correlated .3.

7. An item measuring "responsiveness," "how often the city council responded favorable to the spending preferences of the participant in the past three years," showed results like those in Figure 2.7, so it is not reproduced here.

8. This regression model was specified similarly to that for mayors' welfare spending preferences in Table 2.1.

9. "Like every other urban leader in this country, Harold Washington has had to operate in a national political climate that Terry Clark and Lorna Crowley Ferguson have identified in their interesting book, *City Money*, as the New Fiscal Populism." For instance, "Washington cut the City's non-safety connected work-force by a startling 15% in his first eight months in office" (Hollander 1986, pp. 178, 179).

References

American Enterprise, The. 1991. "Do You Happen to Be Reading?" (July/August):92.

Atkinson, A. B. 1975. *The Economics of Inequality.* London: Oxford University Press.

Baldridge, John. 1991. "New Social Movements and the Emergence of a New Political Culture." Paper prepared for Workshop in Urban Policy, University of Chicago.

Bartels, C. A. and P. Nijcamp. 1976. "An Empirical Welfare Approach to Regional Income Distributions." *Socioeconomic Planning Sciences* 10:244-63.

Bickford, Adam, Terry Nichols Clark, and Alex Ritchie. 1989. *FA for PC.* Chicago: Urban Innovation Analysis.

Birch, David L. 1979. *The Job Generation Process.* Cambridge, MA: MIT Program on Neighborhood and Regional Change.

Bobo, Lawrence and Franklin D. Gilliam. 1990. "Race, Socio-political Participation, and Black Empowerment." *American Political Science Review* 84:344-93.

———, James H. Johnson, Melvin L. Oliver, James Sidanius, and Camille Zubrinksy. 1992. "Public Opinion Before and After the Spring of Discontent." UCLA Center for the Study of Urban Poverty, Occasional Working Paper Series.

Butts, Paul M. 1992. "Feminist Attitudes and Political Involvement: A Cross National Analysis." Paper prepared for Workshop in Urban Policy, University of Chicago.

Clark, Terry Nichols, ed. 1985, 1986, 1987, 1988. *Research in Urban Policy,* Vols. 1, 2A, 2B, 3. Greenwich, CT: JAI.

———. Forthcoming. "Clientelism, USA." In *Democracy, Clientelism, and Civil Society,* edited by Luis Roniger and Aysa Gunes-Ayata. Boulder, CO: Lynne Rienner.

———, Richard Balme, and Rowan A. Miranda. 1988. "Leadership Patterns in American Cities: Modeling Policy Processes." Prepared for session on Urban Political Leadership, Urban Politics Section Panel, American Political Science Association, Washington, DC, September 3.

——— and Lorna Crowley Ferguson. 1983. *City Money: Political Processes, Fiscal Strain, and Retrenchment.* New York: Columbia University Press.

——— and Vincent Hoffmann-Martinot, eds. Forthcoming. *The New Political Culture.* University of Chicago.

——— and Ronald Inglehart. 1988. "The New Political Culture." Presented to annual meeting of American Political Science Association, Atlanta, Georgia.

——— and Seymour Martin Lipset. 1991. "Are Social Classes Dying?" *International Sociology* 4:397-410.

———, Seymour Martin Lipset and Michael Rempel. 1993. "The Declining Political Significance of Social Class." *International Sociology* 8(3):293-316.

——— and Michael Rempel, eds. Forthcoming. *The Politics of Post-Industrial Societies.* University of Chicago.

Clarke, Susan E., ed. 1989. *Urban Innovation and Autonomy.* Vol. 1, Sage Series in Urban Innovation. Newbury Park, CA: Sage.

Coleman, Richard P. and Lee Rainwater with Kent A. McClelland. 1978. *Social Standing in America.* New York: Basic Books.

Coulter, Philip B. 1989. *Measuring Inequality.* Boulder, CO: Westview.

Dahl, Robert A. 1961. *Who Governs?* New Haven, CT: Yale University Press.

Davis, James A. 1989. "Attitudes Toward Free Speech in Six Countries in the mid 1980s." NORC-GSS Cross National Report 9.

Elazar, Daniel. 1984. *American Federalism.* New York: Harper & Row.

Gans, Herbert. 1962. *The Urban Villagers.* Glencoe, IL: Free Press.
Hajnal, Zoltan and Terry Nichols Clark. Unpublished. "Interest Group Politics: Who Governs and Why?" University of Chicago.
Hollander, Elizabeth L. 1986. "The Administration of New Fiscal Populism." Pp. 177-84 in *Managing Cities.* Vol. 2B, *Research in Urban Policy,* edited by Terry Nichols Clark. Greenwich, CT: JAI.
Hout, Michael, Clem Brooks, and Jeff Manza. 1993. "The Persistence of Classes in Post-industrial Societies." *International Sociology* 8 (3):259-78.
Kincaid, John, ed. 1982. *Political Culture, Public Policy, and the American States.* Philadelphia: Institute for the Study of Human Issues.
Lieberson, Stanley. 1969. "Measuring Population Diversity." *American Sociological Review* 34(December):850-62.
Lipset, Seymour Martin. 1981. *Political Man,* 2nd ed. Baltimore, MD: Johns Hopkins University Press.
———. 1991. "No Third Way: A Comparative Perspective on the Left." Pp. 183-232 in *The Crisis of Leninism and the Decline of the Left,* edited by Daniel Chirot. Seattle, WA, and London: University of Washington Press.
Miller, Alton. 1989. *Harold Washington.* Chicago: Bonus Books.
Nie, Norman, Sidney Verba, Kay L. Schlozman, Henry E. Brady, and June Junn. 1989. "Participation in America." University of Chicago.
Pakulski, Jan. 1993. "The Dying of Class or Marxist Class Theory?" *International Sociology* 8(3):279-92.
Patterson, Orlando. 1977. *Ethnic Chauvinism.* New York: Stein and Day.
Peters, Thomas J. and Robert H. Waterman. 1982. *In Search of Excellence.* New York: Warner Books.
Rempel, Mike. 1992. "Social and Fiscal Liberalism in the U.S." Paper prepared for Workshop in Urban Policy, University of Chicago.
Schumaker, Paul and Allan Cigler. 1989. "Bureaucratic Perceptions of the Municipal Group Universe: 1975 and 1989." Pp. 424-34 in *New Leaders, Parties and Groups,* edited by Harald Baldersheim, Richard Balme, Terry Nichols Clark, Vincent Hoffmann-Martinot, and Hakan Magnusson. Paris and Bordeaux: CERVEL.
Schuman, Howard, Charlotte G. Steeh, and Lawrence Bobo. 1988. *Racial Attitudes in America.* Cambridge, MA: Harvard University Press.
Schwartz, Joseph and Christopher Winship. 1979. "The Welfare Approach to Measuring Inequality." In *Sociological Methodology, 1980,* edited by Karl F. Scheussler. San Francisco: Jossey-Bass.
Smith, Tom W. 1985. "Atop a Liberal Plateau?" Pp. 245-58 in Clark 1985.
Stein, Robert M. 1985. *Fiscal Austerity and Urban Innovation Project Codebook.* Department of Political Science, Rice University.
Theil, Henri. 1967. *Economics and Information Theory.* Amsterdam: New Holland.
Tillman, Dorothy. 1986. "Movement Activism From Martin Luther King to Harold Washington." Pp. 155-58 in Clark 1986.
Vanneman, Reeve and Lynn Weber. 1987. *The American Perception of Class.* Philadelphia: Temple University Press.
Ward, Conor and Andrew Greeley. Unpublished. "'Development' and Tolerance: The Case of Ireland." NORC.
Weil, Frederick. 1985. "The Variable Effects of Education on Liberal Attitudes." *American Sociological Review* 50:458-74.
Wilson, William Julius. 1978. *The Declining Significance of Race.* Chicago: University of Chicago Press.

3

Containing Cleavages

Parties and Other Hierarchies

Rowan A. Miranda

Introduction

In recent years, a considerable amount of scholarly attention has been dedicated to the study of fiscal strain (see Levine 1980; Levine, Rubin, and Wolohojian 1981; Clark and Ferguson 1983; Ladd and Yinger 1989). The fiscal distress of numerous large cities has been blamed on urban economic restructuring, economic crises induced by advanced capitalism, deconcentration of population and industry, declining intergovernmental aid, and rising citizen demands for services. Yet, although potentially more acute, the fiscal crises currently facing U.S. cities are not particularly unique. For example, Shefter (1985) notes that New York City faced such crises repeatedly during the past century. Although many studies of fiscal strain cogently make the case for economic constraint explanations, such explanations remain insufficient for reasons outlined below (Wong 1990).[1]

Changes in national and international economic systems can certainly lead to greater fiscal stress among cities. At the same time, as Clark and Ferguson (1983) show, the extent of fiscal stress can vary across cities. For example, at a time when the mayor of New York attributed the city's financial problems to a "Northeast Syndrome," other cities such as Chicago and Pittsburgh faced similar economic circumstances, but avoided major fiscal problems (Clark 1985). The differences between cities in their ability to cope with fiscal stress suggest the potential for alternative explanations from those that have been put forth. This study examines whether the institutional and organizational capacity provided by some political arrangements enables city officials to better cope with fiscal strain and thus avoid fiscal crisis. Alternatively stated, *Do governing crises induce fiscal crises?*

Although there are many political explanations for fiscal strain, only a prominent subset of them is examined here. Taken together, these explanations constitute the "strong party organization" (SPO) theory of urban fiscal politics (Miranda 1993). Because there are several variations of this theory, some oversimplification is unavoidable. The SPO theory states that *cities governed by strong party organizations are more successful in imposing fiscal discipline than cities without such governing arrangements because centralized parties are less responsive to interest group demands for spending.*

The SPO theory is similar to theories of political coalitions in the public choice literature. Parties are organized by elites and voters. Politicians governing under SPOs follow a strategy akin to the minimum winning coalition argument made by Riker (1962). SPOs spend to the extent necessary to maintain the coalition, in contrast to the expectations of the Downsian vote maximization or a "rainbow coalition" strategy of reaching out to numerous ethnic and special interest groups (Erie 1988). The SPO thesis predicts that the latter strategy increases public spending.

The demise of party organizations was heralded for much of the 20th century because of their associated corruption, inefficiency in service delivery, and bias toward certain ethnic groups (Steffens [1903] 1966; Handlin 1951; Hofstader 1955). Subsequently, many have lamented the decline of party politics by arguing that reform institutions promote an upper class bias in public policy (Hawley, 1971; Yates 1977; Welch and Bledsoe, 1988). More than two decades ago, Lowi (1967, p. 86) stated that the decline of the machine and "the triumph of Reform really ends in a paradox," with cities being "well-run but ungoverned."

My principal aim in this study is to outline and test the SPO theory of urban fiscal politics. In my fiscal history of Chicago, mixed support was found for the SPO thesis (Miranda 1993). The political machine may have enabled Chicago's "Boss" Mayor Richard J. Daley to adopt fiscally conservative policies; postmachine mayors, however, were unable to implement expansionary fiscal policies. I overcome limitations imposed by the case study by using data from the Permanent Community Sample (PCS) and the Fiscal Austerity and Urban Innovation (FAUI) Project. The impact of SPOs on the functional responsibility, expenditures, and public employment of city governments is examined over the 1960-1985 period.

I. Defining Strong Party Organizations

Although local parties have weakened or have been replaced altogether by reform institutions during this century, they always varied in strength from situations in which candidates used them merely as labels to stronger organizations such as the classic political machine. SPO building has also varied among cities (DiGaetano 1988). When New York City, which had a machine for decades, was entering an era of reform, Chicago's political machine was being assembled for the first time (Lowi 1967, p. 84). In this study, usage of the term SPO parallels Mayhew's (1986, p. 19) definition of a traditional party organization (TPO).

Mayhew outlined five distinguishing characteristics of these TPOs:

1. *Substantial autonomy.* TPOs have a power base independent of organizations outside the electoral arena such as labor unions and corporations. This gives TPOs leverage in dealing with these organizations.
2. *Long duration.* The life spans of TPOs are measurable "in decades or generations rather than in months or years" (Mayhew 1986, p. 19). Many of these organizations can survive changes in elected political leadership.
3. *Hierarchical organization.* The internal structure of TPOs is hierarchical, including the promotion of candidates. Many TPOs are also named after the "boss."
4. *Nominations for wide range of organizations.* TPOs frequently nominate and slate candidates for municipal, county, state, and congressional offices.
5. *Use of material over purposive incentives.* TPOs rely more on material inducements than purposive or ideological issues to gain support.

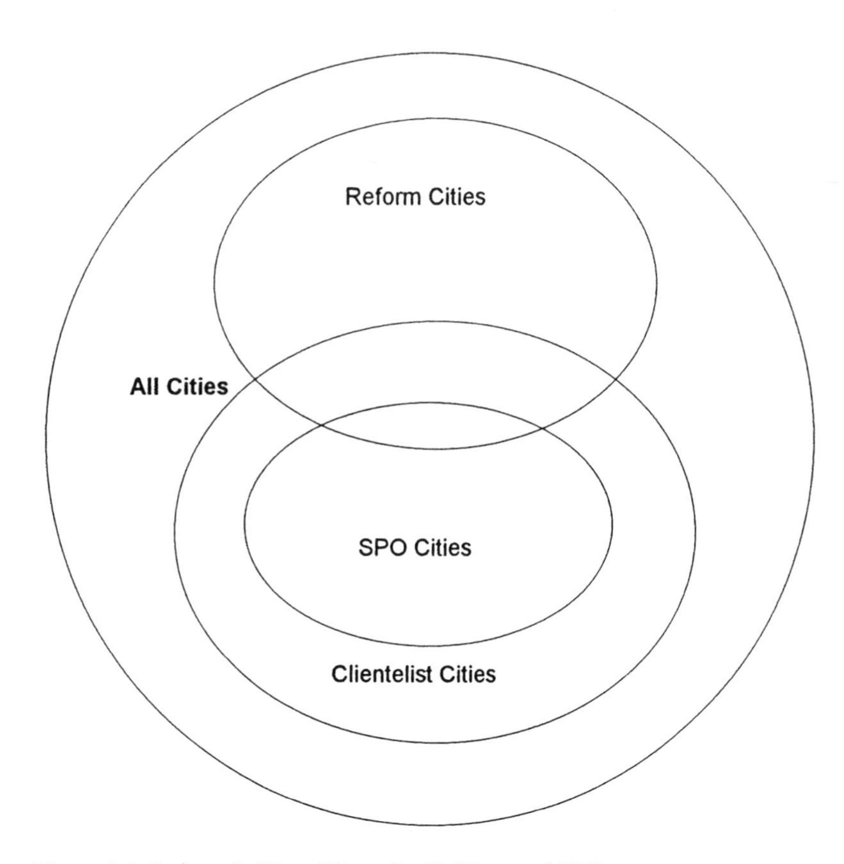

Figure 3.1. Reform Politics, Clientelist Politics, and SPOs

> Patronage in employment and contracts are the two main types of material inducements.[2]

Are all TPOs machines? Mayhew (1986, p. 203) argues that machines are "the strong subspecies of traditional organization." Thus, all machines are TPOs but not vice versa (Figure 3.1). Most machines have had an "overall" control over governments at the city, county, and sometimes state levels. Control of TPOs (and SPOs) is frequently more circumscribed. The "strength" of parties as discussed here is viewed as a continuum with nonpartisan cities and those with weak, factionalized parties at one end and political machines at the other.

Unfortunately, writers have combined analytically distinct terms such as "parties,""bosses,""strong party organizations,""machine politics,""corruption,""patronage," and"centralization." As Wolfinger (1974, p. 99) notes,"the result is to mistake isolated events for historical trends and to propose false tests about the current level of machine politics." In his study of New Haven, Wolfinger distinguished machine politics from political machines:

> *Machine politics* is the manipulation of certain incentives to partisan political participation: favoritism based on political criteria in personnel decisions, contracting, and administration of laws. A *political machine* is an organization that practices machine politics, that is, that attracts and directs its members primarily by means of these incentives. . . . Machine politics (patronage incentives) need not produce centralized organizations *at the city level or higher;* the highest level at which cohesive party organizations are found may be the ward, assembly district, or some other geographic subdivision. Thus for purposes of citywide decision making a party may be fragmented despite the prevalence of machine politics. Patronage and other such incentives, then, are by no means identical to hierarchical party structure [emphasis in original]. (1974, pp. 100-101)

Although many have recited Wolfinger's distinction, in no study has anyone attempted to discriminate between the fiscal policy implications of factionalized clientelist politics and SPO control in cities.

The two dominant groups of theories about political parties and public policy focus on *party competition* or *party ideology.*[3] Hibbs (1987), Wilensky (1975), and Castles and McKinlay (1979) consider the impact of *party ideology* on public policies ranging from inflation and unemployment to the size of the public sector (at the national level). V. O. Key motivated a series of studies seeking to establish whether *party competition* "bids up" state expenditures (Lockard 1963). The SPO theory incorporates aspects of the party competition theory, but not party ideology theory. However, centralization of power and a reliable electoral base are the main features of the SPO theory. Thus, the "party" aspect of the SPO theory is sufficient, but not necessary: any organization that is electorally stable and centralized is predicted to enhance fiscal control of the governing coalition. For example, control of city government in Dallas by the Citizens Charter Association (CCA) seems to have a function similar to that attributed to SPOs—regulating interest group demands (Elkin 1987). Figure 3.2

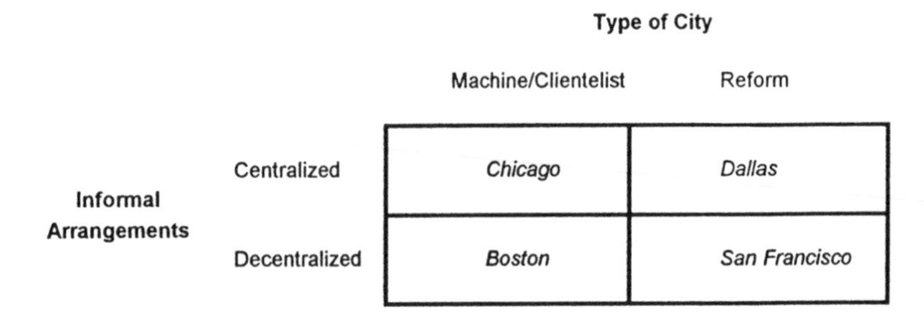

Figure 3.2. Centralization and Decentralization in Urban Governance

provides examples of cities with different governing arrangements. However, because measures of "strong party-like" organizations are not readily available, their impact is not examined. Hereafter, to simplify terminology, machine politics is referred to as "clientelist politics" and political machines as "SPOs."

II. The SPO Theory of Urban Fiscal Politics

Merton (1968, p. 73) stated that the key structural function of the boss "is to organize, centralize and maintain in good working condition the scattered fragments of power." The thrust of Merton's discussion is implicit in Banfield's (1961) *Political Influence.* According to Banfield, a large metropolitan area like Chicago could only be governed if a mayor overcomes *formal decentralization* by *informal centralization.* "By far the most important mechanism through which this is done is the political party or machine" (Banfield 1961, p. 237).

SPOs, like all organizations, have "maintenance" needs (Wilson 1973). The maintenance needs of the party organization can motivate its leaders to sacrifice short-term goals (e.g., jobs and patronage) for long-term or "maintenance"-oriented ones, such as continuity and having voter support in the areas outside the central city when it is necessary to do so. "When every interest has a real chance of affecting an outcome by asserting itself vigorously, incitement to controversy is strong" (Banfield 1961, p. 258). By *informally* centralizing authority through the party organization, interest groups are weakened, and the SPO decides which groups get what from government. A reliable electoral base provides SPOs fewer incentives to be responsive to new interest groups. Thus, as Mayhew hypothesized for states with TPOs, cities with SPOs are expected to have smaller "public economies."

In *The Management of Big Cities,* Rogers (1971) analyzed urban problems in three cities and argued that an important aspect of city governance is the presence of "integrative institutions." Political parties are one of these integrative institutions. Because the reform movement eliminated the machine in New York City, Rogers describes that city's politics as "pluralism run wild." Cole (1974, p. 58) similarly stated that "the decline of the urban machine has left an urban mayor with few (if any) vehicles for the amalgamation of sufficient political power with which to confront contemporary municipal problems."

Shefter (1985) described nearly a century of New York City politics as a "machine/reform dialectic." Because "political crises often manifest themselves as fiscal crises," Shefter argued that such crises "should be regarded not as aberrations but as an integral part of urban politics" (1985, p. 224). A number of imperatives face big-city mayors such as winning votes, promoting economic development, preserving public order, and maintaining the city's credit rating (Elkin 1987). In an effort to win votes and contain social conflicts, city officials are tempted to enact programs faster than they can increase revenues. When downturns in the national economy or structural changes in the local economy occur, "it may be difficult for city officials to keep aloft all the balls they juggle" (Shefter 1985, p. 10). Fiscal crisis follows. Implicit in Shefter's study is the view that some political arrangements provide the organizational capacity for maintaining fiscal control and others promote fiscal strain.

New York City's recent fiscal problems resemble those it has faced repeatedly for more than a century. In the 1860s, Boss Tweed won the support of the city's development community as well as unskilled Irish laborers, which he used to establish Tammany Hall. Because Tweed did not command a disciplined party organization, he embarked on theft and bribery to increase support for the ring's policies and programs (Shefter, 1985, p. 17). Tweed's practices were motivated by the desire to make a "killing in politics" and the need to "hold together a heterogeneous coalition of social groups in the absence of a party apparatus that could reliably produce majorities in the city's electoral arena and policymaking institutions" (1985, p. 18). After the Tweed Ring was toppled in 1871, John Kelley began to centralize and strengthen the Tammany organization. Centralization of the organization meant that "patronage would be distributed through the organization rather than being allocated to individual politicians or public officials who would then distribute it to their personal supporters" (1985, p. 19). Nearly a century later, Richard J. Daley similarly centralized the local government authorities in Chicago by being an elected mayor and retaining chairmanship of the Cook County Democratic Party organization (Miranda 1993).

Reform coalitions and the machine exchanged control of New York City on numerous occasions after 1894. The 1975 fiscal crisis can be traced to changes in the city's political structure during the 1960s that altered the extent of responsiveness of mayors to groups:

> During the decade and a half following World War II, an unusually large number of expenditure-demanding groups were active in New York politics, but because the public official through whom they characteristically obtained access to City Hall—the mayor—was also beholden to the city's regular Democratic party organizations, which represented tax-conscious homeowners and small businessmen, their ability to influence municipal fiscal priorities was tightly constrained. . . . Mayor Wagner's break with Carmine DeSapio and his associates in 1961 and the election of a fusion mayor in 1965 altered the pattern of relations among the major forces in New York politics, and weakened this constraint on budgetary inflation. (Shefter 1985, p. 254)

Although fiscal restraint was internally imposed before the 1975 crisis, it was externally imposed afterward (e.g., via the Municipal Assistance Corporation and the Emergency Financial Control Board).

Fuchs (1992) applies the SPO theory to explain differences in the fiscal policy patterns of New York and Chicago. Chicago fares as well as New York on some measures of economic hardship and worse on others. Yet Chicago had fewer fiscal problems. What explains the differences in spending patterns? Fuchs argues that a mayor's "capacity to centralize and control the budgetary process and to limit the demands on the system are the two most important elements of the fiscal policy process that affect fiscal stability" (1992, p. 3). Specifically, Chicago's Daley machine avoided fiscal insolvency:

> because [Chicago's] leaders were able to deflect interest group demands for increased spending. Why were New York mayors unable, unlike Chicago mayors, to say no to interest group demands and to

> control the budget decisions with an eye toward fiscal instability? . . . New York mayors were unable to centralize authority over the budgetary process after the destruction of the Democratic machine. Without a strong party organization New York mayors relied on interest group support to create winning electoral coalitions; their support came with a fiscal price tag. (Fuchs 1992, p. 6)

In *Rainbow's End,* Erie (1988) examines political machines in eight heavily Irish cities. He demonstrates that the fiscal policy patterns adopted by machines were influenced by changes at the national and state levels. As one would expect from a careful historical study spanning 150 years, Erie's masterful portrayal of machine politics is considerably more complex than the SPO theory discussed thus far.

Much of Erie's discussion, however, is generally consistent with the theory. He cogently argues against the "rainbow theory of politics," which states that the Irish "used a political route to travel from rags to riches, capturing patronage-laden machines and turning public employment into an Irish preserve" (1988, p. 1). He demonstrates that the supply of resources available to machines was too small to enable the Irish to accomplish this. "Bosses rarely possessed the cornucopia of benefits" suggested by the rainbow theory (p. 210).

Consistent with the SPO theory, leaders of the entrenched machines were "selective mobilizers" who "jealously guarded" municipal resources and did not use them to assimilate later arriving Southern and Eastern Europeans and blacks (p. 10).

> Mature machines were one-party regimes lacking the political incentive to mobilize the second-wave of immigrants. The Irish Democratic bosses had already constructed winning electoral coalitions among early-arriving ethnic groups. The newcomers' political assimilation would only encourage demands for a redistribution of power and patronage. (Erie 1988, p. 244)

Erie separates the "embryonic" stage of machine building—in which leaders are motivated to pursue expansionary fiscal policies to build support—with the consolidated phase, in which a stable electoral base leads leaders to be less motivated to do so. Post-New Deal machines developed a strain of fiscal conservatism as their Irish constituencies moved from being service-demanding renters to tax-paying homeowners. This fiscal conservatism was also reinforced by a relaxation in demands for municipal jobs by the machine's core constituency. To secure middle-class votes, entrenched machines:

> had to devise a much different menu of policies. Middle-class voters were homeowners, sensitive to tax increases and less desirous of patronage and welfare services. Middle-class voters demanded low taxes and home-owners services such as garbage collection and street cleaning. The longevity of Irish machines of Chicago, Albany, and Pittsburgh well into the post-World War II era is attributable to their ability to shift from working-class to middle-class policies for white ethnics while piggy-backing welfare-state programs for blacks and Hispanics. (1988, p. 15)

Even if leaders had the incentive to pursue expansionary fiscal policies, hostile Republican governors imposed laws that prevented them from doing so. Erie discusses the role of intergovernmental alliances that successful machine leaders built in an effort to expand the supply of resources. Finally, during the post-New Deal era, instead of political patronage, the principal concern of the machine's traditional constituencies was the preservation of white neighbor-hoods (Kleppner 1985; Erie 1988).

In summary, the history of machine politics discussed in Erie (1988) is fairly consistent with the SPO theory. Although the SPO theory attributes fiscal conservatism to *structural* factors, Erie's analysis attrib-utes it to multiple reasons such as intergovernmental relations, citizen preferences, and alliances with business. In either case, both the SPO and Erie's arguments imply that the welfare state role that Merton, Dahl, and others attributed to political machines is largely unwarranted. At best, machines enabled immigrants to post modest gains through pa-tronage. But Erie (1988, p. 242) raises the question whether these gains actually retarded the economic progress of the Irish as a group by delaying their assimilation into the middle class.

Implicit in some statements of the theory is that SPOs enable political leaders to manage conflicting interest group demands (Shefter 1985; Mayhew 1986; Fuchs 1992). Fiscal conservatism is possible under times of fiscal stress because SPOs can be less responsive to prospending interest groups. Jones (1983) makes similar observa-tions on the relation of party strength and group activity:

> The power of interest groups in local politics is enhanced where politi-cal parties are inactive. This is the case for three reasons. First, where

> parties perform their usual functions of recruitment for office and mobilization of voters in elections, the electoral power of interest groups is weak. . . . Second, in some cases parties and groups compete directly, and the competition reduces the power of groups. . . . Third, interest groups tend to be weaker where parties are strong because . . . parties must create electoral coalitions to govern, they have a *stake in compromise.* Such enforced compromise limits the power of groups in the political process. (p. 148)

The SPO enabled the Daley machine in Chicago to forestall the redistributive demands of minorities and municipal employees for nearly two decades (Miranda 1993). Of course, there is nothing desirable about unresponsiveness to groups from the standpoint of popular control of government (Elkin 1987). Yet, as the case of New York City reminds us, political leaders are frequently tempted to be responsive to numerous groups to maintain power. However, their inability to *balance the demands on the political system with the available supply of resources may induce a fiscal crisis that in turn unravels their governing coalitions.*

III. Testing the SPO Theory

Hypotheses

Five hypotheses that state the impact of SPOs on policy outputs are discussed below.

H1: Cities with a tradition of clientelism have a larger scope of functional responsibility.

Several studies have attempted to explain the level of functional responsibilities that cities face. Dye and Garcia (1978) found that older, larger, northeastern cities have broader responsibilities. Liebert (1976) and Stein (1990) found that functional responsibility is positively related to city age. Do cities with a clientelistic political culture have greater functional responsibilities?

Cities with a history of machine politics are likely to be burdened with more service responsibilities because this enables machine politicians to increase opportunities for patronage and provide ethnic voters with more services. Peterson (1981, p. 206) states that once a city "assumes responsibility for a public function, it cannot easily discard it." By contrast, younger postindustrial cities are expected to have a minimalist scope of service responsibilities (Stein 1990).

H2: Cities with a tradition of clientelism have higher expenditure, employment, and compensation levels.

H3: SPO cities have lower expenditure levels.

H4: SPO cities have higher employment levels and lower compensation levels.

In "The Irish Ethic and the Spirit of Patronage," Clark (1975b) examined the impact of Irish machine politics on city spending over a period spanning nearly a century (1880-1968). Cities with a higher proportion of Irish residents had greater expenditures for police and fire. Clark (1975b, p. 346) conducted a battery of tests for potential spuriousness of *percent Irish* with other variables and found the positive relationship still held. He concluded that the Irish ethic:

> legitimated and reinforced the spirit of patronage politics: a style of leadership distinguished by its use of separable goods, resources allocable to distinct individuals and social sectors. The basic resources exchanged in patronage politics are favors, votes, and government jobs. Such resources helped the Irish succeed dramatically as political candidates. Favors dispensed by political candidates "got out the vote." Campaigners were supported by government jobs, which Irish officials increased. This in turn raised municipal expenditures. (p. 305)

Do Clark's findings—that *clientelism* leads to higher expenditures—conflict with the SPO thesis? In short, no. The SPO thesis does not dispute that clientelist politics leads to higher spending. Erie's study suggests that this is likely to occur during the embryonic stage of machine building when leaders are attempting to gain power. It is also similar to V. O. Key's idea that party competition, or factionalism, increases taxes and spending. What the SPO thesis suggests is that SPOs structurally induce fiscal conservatism because reliable electoral majorities allow them to be less responsive to prospending interest groups. Thus, although Chicago and Boston both have an Irish political heritage, the fact that the former city developed an SPO whereas Boston "never rose above factional ward politics," has fiscal policy consequences (Erie 1988, p. 9).

Because of their emphasis on separable goods (Clark 1975b), clientelist politics cities are expected to increase expenditures and employment in comparison to cities with reform institutions.[4] The presence of an SPO, however, is expected to be negatively associated with spending levels. This follows from the more abstract proposition "centralization encourages public goods, but decentralization generates separable goods" (Clark 1975a). Because SPOs practice constrained clientelist politics, the imperatives facing SPO leadership may lead them to adopt higher levels of municipal employment to meet patronage demands. However, to resolve the contradictions posed by fiscal conservatism *and* higher municipal employment, SPOs adopt lower compensation levels. After all, vote-maximizing machine politicians would want to squeeze out as many jobs as possible from the same level of resources, because votes are a fungible "currency."

H5: SPO cities are associated with lower expenditure growth. Cities with machine politics are associated with higher expenditure growth.

Fuchs's (1992) study of Chicago and New York focuses on the ability of a mayor to *control city spending. An analysis of current expenditure and employment levels* tells only part of the story. A history of machine politics may lead some cities to spend more than others, especially during the "embryonic phase" (Erie 1988), but the presence of an SPO should also keep the rate of expenditure *growth* lower than for other cities. New York City's fiscal problems resulted from the inability of mayors to keep spending in pace with changes in the private sector resources of that city. Ultimately, it is not absolute spending levels that cause fiscal stress, but the inability to control spending growth. Thus, a stronger test of the SPO theory would demonstrate that the *rate of expenditure growth* is lower in SPO cities than in cities with machine politics.

Data

Demographic, expenditure, and employment data for this study were obtained from the Permanent Community Sample (PCS), the County and City Data Book, and the U.S. Census Bureau's City

Government Finance series. The *measure for SPOs* is constructed by coding individual cities based on David Mayhew's (1986) *Placing Parties in American Politics.* The proxy for *clientelist/machine politics* is percentage of Irish-stock residents. The *index of functional performance* is used to measure the scope of service responsibility. It was computed by using the method discussed in *City Money* (Clark and Ferguson 1983, p. 314).

Motivated by Lazarsfeld's (1959) discussion of the interchangeability of indexes, data from the Fiscal Austerity Urban Innovation (FAUI) Project survey were used to compute an *alternative SPO index.* The alternative index is admittedly more a measure of the local party activity or salience than strength. By party organization strength, the importance of the political party to elected officials (particularly the mayor) is actually being measured. The index was constructed by adding mayor's responses to three questions in the FAUI survey: activity of party in the campaign, mention of party affiliation in the campaign, and how often the mayor met with local party officials.[5] Although this index is less than ideal, it enables comparison of the findings against the SPO measure coded from the Mayhew study.

Findings

Table 3.1 presents results from a regression model explaining the level of functional responsibilities across cities. The model controls for the major explanations found to be important in past studies, namely, city age and region. Consistent with H1, cities with a tradition of machine politics bore a larger responsibility in service provision. Although city age is positively associated with functional responsibility, unlike the case in past studies, the Northeast region is found to have a statistically insignificant impact on functional responsibility. This finding that machine politics increases the scope of functional responsibility supports H1. It is also consistent with Clark and Ferguson's (1983) claim that local political culture, not the "Northeast Syndrome," explains service burdens. Consistent with the theory, cities governed by SPOs were no more likely to be responsible for a broader scope of services than other cities.

Table 3.2 presents results on expenditures, employment, and compensation. Consistent with H2, cities with a tradition of machine politics have *higher* levels of expenditures, employment, and compensation. There is also partial support for H3: cities governed by

TABLE 3.1 Explaining the Scope of Urban Functional Responsibility

Independent Variables	*B*	*T*	*Sig.T*
Northeast Dummy	0.37	0.90	.37
Population	0.20	1.68	.10*
Organization of Municipal Employees	0.74	1.05	.30
Percent Black	0.18	1.39	.17
City Age	0.68	2.37	.02***
Percent Irish Residents	0.68	2.14	.04***
Strong Party Organization Dummy	−0.13	−0.47	.64
Constant	−1.14	−1.23	.22
R Square	.62		
Adjusted *R* Square	.58		
Number of Cities	63		

NOTE: * = $p < .10$; ** = $p < .05$; *** = $p < .01$.

SPOs have *lower* expenditures. Although it does not necessarily contradict the SPO theory, the results provide mixed support for H4. Cities governed by SPOs have lower compensation levels. However, although the coefficient shows the correct sign, the impact of SPOs on employment is statistically insignificant. Fuchs (1992) found that the Chicago machine consistently increased employment *after* elections to reward machine supporters. The insignificant impact of SPOs on employment could result from an inability of our analysis to separate out the timing of elections. Nevertheless, overall the analysis indicates that there are distinctive fiscal policy consequences of clientelist politics and SPOs. The latter are generally fiscally conservative whereas the cities with clientelist politics *alone* adopt expansionary fiscal policies.

Are SPOs able to better control the *growth* of city government spending than cities without such governing arrangements? Two sets of analyses were conducted. Table 3.3 shows results for the 1960-1970 period. Cities with a tradition of machine politics were more likely to increase spending and compensation. Cities with SPOs increased employment during this period, but decreased compensation—a pattern consistent with H4. However, SPOs did not have a statistically significant impact on the overall growth of city government expenditures.

TABLE 3.2 Determinants of Expenditure, Compensation, and Employment Levels, 1970

	Common Functions Expenditures per Capita			*Personnel Expenditures per Capita*			*Municipal Employment per Capita*		
Independent Variables	*B*	*T*	*Sig.T*	*B*	*T*	*Sig.T*	*B*	*T*	*Sig.T*
Functional Performance Index	—	—	—	0.16	2.78	.01***	0.18	3.30	.00***
Median Income	0.27	1.19	.24	−0.04	−0.11	.91	−1.11	−3.30	.00***
Intergovernmental Aid	0.18	5.68	.00***	0.28	5.19	.00***	0.22	4.30	.00***
Organization of Municipal Employees	0.03	0.13	.90	0.09	0.32	.75	−0.45	−1.61	.11
Percent Black	0.11	3.50	.00***	0.10	2.04	.05**	0.05	1.02	.31
Percent Irish Residents	0.28	4.26	.00***	0.36	3.32	.00***	0.31	3.05	.00***
Strong Party Organization Dummy	−0.15	−2.15	.04**	−0.22	−2.13	.04**	−0.09	−0.90	.37
Constant	1.02	0.49	.63	2.97	0.89	.38	11.25	3.58	.00***
R Square		.70			.80			.79	
Adjusted *R* Square		.66			.77			.76	
Number of Cities		63.00			63.00			63.00	

NOTE: * = $p < .10$; ** = $p < .05$; *** = $p < .01$.

The latter finding is potentially a result of the period examined. The fiscal condition of many large cities during the 1960s was quite good, in comparison to subsequent decades. Increases in intergovernmental aid allowed cities to increase expenditures at a time when the economy was growing. There is evidence in studies by Erie (1988) and Fuchs (1992) that SPOs adopt more fiscally conservative policies only when they have to—in times of fiscal stress. For this reason, the 1980-1985 period, in which cities lost intergovernmental aid and the national economy was sluggish, was also examined.

Table 3.4 shows the impact of two measures of SPOs—the Mayhew SPO variable and the FAUI SPO index—on common functions and general expenditures. Irrespective of the measure, the findings are generally consistent. Although the effects of the respective measures are small, they are statistically significant. In Model A, cities governed by SPOs had a lower growth of general and common function expenditures. In Model B, although the sign for general expenditures is negative, the impact of the alternative SPO index is significant only for common function expenditures. On the whole, the findings for the 1980-1985 period should be interpreted more tentatively, as Mayhew's TPO score is less valid for this period.

IV. Discussion

In this study, I attempted to test the SPO theory of urban fiscal politics and examined the fiscal policy consequences of Wolfinger's distinction between political machines (i.e., SPOs) and machine politics (i.e., clientelism). Overall, the results generally support the SPO theory. Cities governed by SPOs have lower expenditure *levels* and lower expenditure *growth*, especially during difficult economic times. In particular, the evidence suggests that clientelist cities generally adopt expansionary fiscal policies in comparison with their reform counterparts unless they develop an SPO. Both centralization and the reliable electoral majorities typical of SPOs allow political leaders to impose fiscal discipline. The findings presented in this study do marshal a modest level of support for the SPO theory. Further, the findings imply that scholars studying big-city politics should carefully consider Wolfinger's (1965, 1974) distinction between machine politics (i.e., clientelist politics) and political machines.

TABLE 3.3 Sources of Public Sector Growth, 1960-1970

Independent Variables	*Change in Expenditures*			*Change in Employee Compensation*			*Change in Municipal Employment*		
	B	*T*	*Sig.T*	*B*	*T*	*Sig.T*	*B*	*T*	*Sig.T*
Population Change	–0.68	–2.38	.02**	0.22	0.76	.45	3.86	4.19	.00***
Change in Median Income	–0.52	–1.31	.20	–0.02	–0.06	.95	–0.41	–0.31	.76
Change in Intergovernmental Aid	0.15	3.34	.00***	0.05	0.97	.34	0.12	0.84	.40
Organization of Municipal Employees	0.12	0.42	.68	0.72	2.49	.02**	–0.11	–0.12	.91
Change in Percent Black	0.06	2.02	.05**	0.05	1.57	.12	–0.16	–1.69	.10*
Percent Irish Residents	0.17	1.82	.08*	0.27	2.91	.01***	–0.24	–0.80	.43
Strong Party Organization Dummy	–0.07	–0.68	.50	–0.24	–2.30	.03**	0.91	2.74	.01***
Constant	9.06	6.12	.00***	2.28	1.51	.14	–13.3	–2.78	.01***
R Square		.49			.39			.39	
Adjusted *R* Square		.41			.29			.29	
Number of Cities		51.00			51.00			51.00	

NOTE: * = $p < .10$; ** = $p < .05$; *** = $p < .01$.

TABLE 3.4 Sources of Public Sector Growth, 1980-1985

	Model A						*Model B*					
	Change in per Capita General Expenditures			*Change in per Capita Common Functions Expenditures*			*Change in per Capita General Expenditures*			*Change in per capita Common Functions Expenditures*		
Independent Variables	*B*	*T*	*Sig.T*	*B*	*T*	*Sig.T*	*B*	*T*	*Sig.T*	*B*	*T*	*Sig.T*
Change in Intergovernmental Aid	0.28	5.31	.00***	0.21	3.94	.00***	0.29	3.17	.00***	0.18	1.86	.07*
Change in Median Income	−0.33	−1.74	.08*	−0.29	−1.49	.14	0.04	0.10	.92	0.03	0.06	.95
Organization of Municipal Employees	−0.13	−2.26	.03**	−0.11	−1.79	.07*	−0.05	−0.50	.62	−0.02	−0.19	.85
Percent Black	−0.20	−2.26	.03**	−0.11	−1.29	.20	−0.25	−1.56	.13	−0.12	−0.71	.48
Strong Party Organization Dummy	−0.06	−2.08	.04**	−0.07	−2.41	.02**	—	—	—	—	—	—
FAUI SPO Index	—	—	—	—	—	—	−0.04	−1.03	.31	−0.12	−2.65	.01***
Constant	0.73	7.98	.00***	0.76	8.23	.00***	0.60	3.36	.00***	0.73	3.81	.00***
R Square		.25			.17			.28			.25	
Adjusted *R* Square		.22			.15			.20			.19	
Number of Cities		170.00			170.00			57.00			57.00	

NOTE: * = $p < .10$; ** = $p < .05$; *** = $p < .01$.

How do our findings compare to past studies? This study generally extends support to arguments made by Shefter (1985), Mayhew (1986), Erie (1988), and Fuchs (1992). First, the findings support Erie's argument against the rainbow theory of politics. Erie's focus on the available supply of municipal resources led him to conclude machine politicians did not deliver benefits on a large scale to their constituents. At best, they were successful at rewarding their constituents with modest patronage, but even here the compensation may have been lower than that for employees in other cities.

Second, consistent with Mayhew's findings for U.S. states, traditional party organizations are associated with smaller budgets. Mayhew attributes the lower spending associated with TPOs to the distinctive political pressures they face:

> What used to be called "pressure politics"—a variety of free-wheeling pressure group activity—seems to have had a low incidence in organization states. . . [because of] their party organizations' incentive structures, their relative lack of openness to interest groups, low "issue density" in their electoral politics, their relative inhospitality to bureaucracy, and political cultures grown up around their electoral and governmental processes. (1986, p. 12)

The findings in this study also imply that factionalism (what Shefter 1976 labels as "rapacious individualism") even among cities leads to increased spending.

Third, both Shefter (1985) and Fuchs (1992) argue that fiscal crises are potentially larger signals of a governing crisis. Ironically, this suggests that to be fiscally sound, large cities need to be undemocratic. For example, whereas SPOs like the Daley machine were able to remain fiscally conservative for decades, they did this at the expense of blacks and the poor. By contrast, New York's politicians were more responsive to their constituencies, which led the city to the brink of bankruptcy and toppled its mayors.

Finally, the emphasis on *informal arrangements* in the SPO theory is consistent with the approach taken in recent work on the governance of metropolitan areas. For example, in his study of the public transit system of the San Francisco Bay Area, Chisholm (1989) argues that a formally fragmented patchwork of transit authorities can be coordinated effectively through informal arrangements. Thus, formal fragmentation should not imply that consolidation is desirable if informal arrangements are effective. With the demise of party organizations, informal networks associated with parties are weakened. Consequently, cities may have mitigated corruption, but increased problems with governance (Yates 1977).

The general decentralization of decision making in U.S. cities in recent years encourages "buck passing" concerning fiscal problems and other "public goods," such as environmental pollution. But as the environmental example suggests, if there is sufficient public concern on an issue, more centralized processes, and even stringent decisions, can become popular. Strong parties may not be the wave of the future, but the SPO is just one form of centralization. Recognizing the public goods and fiscal benefits that can accrue from centralization (Clark 1975a) can lead one to explore centralized alternatives (e.g., the city manager, Emergency Financial Control Board [as in New York], Board of Fiscal Overseers [as in San Diego]) that retain some of the benefits of the traditional strong party organization in a changing sociopolitical environment.

Notes

1. In this study, the term *fiscal strain* refers to city spending levels that exceed the capacity of the private sector resources base (Clark and Ferguson 1983). In situations of extreme fiscal strain, a *fiscal crisis* develops, with cities frequently threatening bankruptcy if intergovernmental aid is unavailable.

2. Wilson (1973) distinguishes material from purposive incentives for political participation. Purposive incentives "are intangible rewards that derive from the sense of satisfaction of having contributed to the attainment of a worthwhile cause." Material incentives "are tangible rewards: money, or things and services readily priced in monetary terms" (pp. 33-34).

3. Studies of urban fiscal policy outputs generally have not found an impact of party ideology on public spending in the United States. For this reason, the party of the mayor is not incorporated as a competing explanation here. Using the FAUI data, each of the regressions did examine the role of both party and the mayor's fiscal liberalism, which did not show statistically significant effects.

4. Numerous studies have examined the influence of *reform institutions* (i.e., council-manager form of government, at-large elections, and nonpartisan elections) on city spending patterns (Lineberry and Fowler 1967; Clark 1968; Lyons 1978; Morgan and Pelissero 1980). Findings from these studies have been generally mixed, although the most carefully conducted studies favor the view that reform institutions have no impact on city spending (Morgan and Pelissero 1980). Because the incidence of reform institutions, to some extent, is the opposite of the variable measuring machine politics, it is not incorporated into the model. Gimpel (1992) finds similar results when his

measure of machine politics and a reformism index were interchanged in predicting expenditures.

5. Two of the three items in the index were measured on a five-point scale and the last item on a four-point scale.

References

Banfield, E. 1961. *Political Influence: A New Theory of Urban Politics*. Toronto: Free Press.

Castles, F. and R. McKinlay. 1979. "Public Welfare Provision in Scandinavia and the Sheer Futility of the Sociological Approach to Politics." *British Journal of Political Science* 9:157-71.

Chisholm, D. 1989. *Coordination Without Hierarchy: Informal Structures in Multiorganizational Systems*. Berkeley: University of California Press.

Clark, T. N. 1968. "Community Structure, Decision-Making, Budget Expenditures and Urban Renewal in 51 American Communities." *American Sociological Review* 33:576-93.

———. 1975a. "Community Power." Pp. 271-296 in *Annual Review of Sociology*, Vol. 1, edited by Alex Inkeles. Palo Alto, CA: Annual Reviews.

———. 1975b. "The Irish Ethic and the Spirit of Patronage." *Ethnicity* 2:327-43.

———. 1985. "Fiscal Strain: How Different are Snow Belt and Sun Belt Cities?" Pp. 253-80 in *The New Urban Reality*, edited by P. Peterson. Washington, DC: Brookings Institution.

——— and L. Ferguson. 1983. *City Money*. New York: Columbia University Press.

Cole, R. 1974. *Citizen Participation and the Urban Policy Process*. Lexington, MA: D. C. Heath.

DiGaetano, A. 1988. "The Rise and Development of Urban Political Machines: An Alternative to Merton's Functional Analysis." *Urban Affairs Quarterly* 24:242-67.

Dye, T. and J. Garcia. 1978. "Structure, Function, and Policy in American Cities." *Urban Affairs Quarterly* 22:151-65.

Elkin, S. 1987. *City and Regime in the American Republic*. Chicago: University of Chicago Press.

Erie, S. 1988. *Rainbow's End*. Berkeley: University of California Press.

Fuchs, E. 1992. *Mayors and Money*. Chicago: University of Chicago Press.

Gimpel, J. 1992. "The Rising Cost of Party Organizational Maintenance." Paper presented at the Annual Meeting of the American Political Science Association, Chicago.

Handlin, O. 1951. *The Uprooted*. New York: Grosset and Dunlap.

Hawley, W. 1971. *Nonpartisan Elections and the Case for Party Politics*. New York: John Wiley.

Hibbs, D. 1987. *The American Political Economy*. Cambridge, MA: Harvard University Press.

Hofstader, R. 1955. *The Age of Reform*. New York: Knopf.

Jones, B. D. 1983. *Governing Urban America*. Boston: Little, Brown.

Kleppner, P. 1985. *Chicago Divided: The Making of a Black Mayor*. DeKalb: Northern Illinois University Press.

Ladd, H. and J. Yinger. 1989. *America's Ailing Cities*. Baltimore, MD: Johns Hopkins University Press.

Lazarsfeld, P. 1959. "Problems in Methodology." In *Sociology Today*, edited by Robert Merton. New York: Basic Books.
Levine, C., ed. 1980. *Managing Fiscal Stress*. Chatham, NJ: Chatham House.
———, I. Rubin, and G. Wolohojian. 1981. *The Politics of Retrenchment*. Beverly Hills, CA: Sage.
Liebert, R. 1976. *Disintegration and Political Action: The Changing Functions of Government in America*. New York: Academic Press.
Lineberry, R. and E. Fowler. 1967. "Reformism and Public Policies in American Cities." *American Political Science Review* 61:701-16.
Lockard, D. 1963. *The Politics of State and Local Government*. New York: Macmillan.
Lowi, T. 1967. "Machine Politics—Old and New." *Public Interest* (Fall):83-92.
Lyons, W. 1978. "Reform and Response in American Cities." *Social Science Quarterly* 59:118-52.
Mayhew, D. 1986. *Placing Parties in American Politics*. Princeton, NJ: Princeton University Press.
Merton, R. 1968. *Social Theory and Social Structure*. New York: Free Press.
Miranda, R. 1993. "Post-Machine Regimes and the Growth of Government." *Urban Affairs Quarterly* 28:397-422.
Morgan, D. and J. Pelissero. 1980. "Urban Policy: Does Structure Matter?" *American Political Science Review* 72:999-1006.
Peterson, P. 1981. *City Limits*. Chicago: University of Chicago Press.
Riker, W. 1962. *The Theory of Political Coalitions*. New Haven, CT: Yale University Press.
Rogers, D. 1971. *The Management of Big Cities*. Beverly Hills, CA: Sage.
Shefter, M. 1976. "The Emergence of the Political Machine: An Alternative View." Pp. 49-71 in *Theoretical Perspectives in Urban Politics*, edited by W. Hawley et al. Englewood, NJ: Prentice-Hall.
———. 1985. *Political Crisis/Fiscal Crisis: The Collapse and Revival of New York City*. New York: Basic Books.
Steffens, L. 1966. *Shame of Cities*. New York: Hill and Wang. Originally published in 1904.
Stein, R. M. 1990. *Urban Alternatives*. Pittsburgh, PA: University of Pittsburgh Press.
U. S. Bureau of the Census. Various years. *City government finances*. Washington, DC: U.S. Government Printing Office.
Welch, S. and T. Bledsoe. 1988. *Urban Reform and Its Consequences*. Chicago: University of Chicago Press.
Wilensky, H. 1975. *The Welfare State and Equality*. Berkeley: University of California Press.
Wilson, J. Q. 1973. *Political Organizations*. New York: Basic Books.
Wolfinger, R. 1965. "Why Political Machines Have Not Withered Away and Other Revisionist Thoughts." *Journal of Politics* 34, 2:365-98.
———. 1974. *The Politics of Progress*. Englewood Cliffs, NJ: Prentice-Hall.
Wong, K. 1990. *City Choices*. Albany: State University of New York Press.
Yates, D. 1977. *The Ungovernable City*. New Haven, CT: Yale University Press.

PART TWO

Making Fundamental Policy Choices

4

The Antigrowth Machine

Can City Governments Control, Limit, or Manage Growth?

Terry Nichols Clark
Edward G. Goetz

GROWTH CONTROL POSES A CHALLENGE to leading urban growth theories, especially Molotch's growth machine, Peterson's developmental policies, and Stone and Elkin's business regimes. Why?

Ecology movements, recycling, and environmental conservation efforts have become important. This is clear from many surveys of citizens and leaders.[1] This chapter is part of a more general effort to describe and interpret these policy shifts. We analyze several data sets to assess what we know about U.S. local government efforts to

AUTHOR'S NOTE: Presented at the annual meeting of the European Consortium for Political Research, Workshop on New Patterns of Local Politics in Europe, Leiden University, Holland, April 2-8, 1993. This chapter emerged from continuing discussions by the authors and others at Roundtable sessions every 6 months for 2 years at the Midwest Political Science Association and American Political Science Association, plus the Annual Urban Politics Conferences in Chicago (1991) and Boulder (1992). Participants at these sessions have exchanged ideas as well as data. Data for this chapter (and most of the book) are available on PC diskette from the authors.

control, limit, or manage growth. These permit us to analyze distinct socioeconomic and political factors that encourage or discourage growth-related movements and policies.

Besides their policy importance, growth and antigrowth policies of city governments are a strategic research site for general urban theories. Conceptually, growth-related movements offer a fundamental challenge to several strands of current urban theorizing, as the theories often imply that local governments should *never* limit growth. Yet this theoretical conclusion is obviously refuted by what we see daily: the visible importance of growth management and ecologically informed policies. Toward an aim of clarity, we identify the general conceptual roots—and limits—of some recent works and introduce alternative theories.

We first review past urban work to specify distinct hypotheses. These hypotheses are then considered with results from four national surveys. Brief case studies of San Francisco, Seattle, Los Angeles, Boulder, and Boca Raton complement the comparative analyses and lead into our general interpretations (see boxed case studies). Chapter Appendix 1 provides technical detail. Chapter Appendix 2 reviews past empirical work on growth controls and management.

Our results seriously question the validity of the capitalism and business regime hypotheses, and point to the importance of citizen preferences, organized groups, and political leaders as important actors.

Past Efforts to Explain Growth and Growth-Limiting Policies

Growth controls, growth limits, and growth management have distinct connotations and multiple subdimensions we consider below, but they share a common policy thrust: to challenge and question the endless expansion of population and jobs. They refer to government policies that seek to direct, shape, or limit in some manner the nongovernmental processes driving growth. The decision "to grow or not to grow" is our main concern in this chapter, which we approach by probing the dynamics of decisions by city governments to limit growth.

What do major lines of urban theorizing posit, or ignore, about growth and growth-limiting policies?

Capitalism and Competition

One set of arguments is in Elkin (1987) and Kantor with David (1988). Like Peterson earlier in *City Limits* (1981), they portray a strong form of the competitive market: *localities must encourage business development* for if they do not, land and local business will decline in value. The competitive market for investment by homeowners and investors drives localities toward constant growth. Cities that do not compete must decline. This national (and international) market competition acutely constrains other possible policies, especially redistribution of wealth toward the disadvantaged.

From this perspective, it is simply "irrational" for a local government not to compete for growth—and even more irrational to limit growth. Yet, despite the popularity of this argument, few studies have sought to consider key components or assess them systematically, especially for growth-limiting policies.

A growing body of work undercuts this capitalism and competition hypothesis. Empirical work even in the "Reagan 1980s" increasingly documented deviations from the strong prodevelopment pattern. Case studies such as Clavel (1986), Clavel and Kleniewski (1990), and Turner (1992) identify specific cities that expanded the scope of "possible" development policies. Selected national urban studies (Clarke and Rich 1985; Goetz 1990) indicate that many local governments violate the logic of progrowth development. Thus, although we do not deny the existence of competitive market forces, we must ask: how can we reconcile these forces with the empirical presence of growth limits by cities? The limits suggest that market competition has been conceptualized in an overly determinist manner in implying that all cities must seek to grow. Because so many governments violate this theory, we need better theory to specify why some governments seek to limit growth.

Strong Business, Growth Machines, and Progrowth Regimes

Although abstractions like "the market" or "capitalist imperatives" are invoked by the first set of theories, a second stresses more direct business leadership. Instead of "the invisible hand," business leaders are the proximate actors. This continues that half of the community power tradition—from the Lynds through Hunter and Domhoff—which held that business dominates government. (Dahl 1961 encouraged the other half to stress more political leadership.) A corollary is that because business wants growth, government must encourage growth. This outlook informs the "growth machine" imagery of Molotch (1976) and many others who generally posit an economic determinism. Their conclusions imply that government is broadly subordinate to business leadership, which seeks continual growth. Clarence Stone's (1989a, 1989b) and Steven Elkin's (1987) discussions of "regime" open a conceptual door for local leaders to redirect impersonal "market forces" and for nonbusiness leaders to emerge. Yet their examples still stress business domination of government, whether direct or subtle, and commitment to growth as civic policy.[2]

Note that Stone's and Elkin's main evidence comes from case studies of two cities that nicely illustrate their points—Atlanta and Dallas. Both cities are nationally distinguished for business boosterism. Although the authors mention contrasting types of "regimes," these get far less attention. Stone's typologies of regimes (e.g., Stone, Orr, and Imbroscio forthcoming) make variations explicit, but the leitmotif remains business.[3]

Although the term "regime" was introduced by Elkin and Stone, they explicitly state that their concept overlaps considerably with such related concepts as leadership, power structure, policy preferences, and political culture. If most regime research has used case studies, its key ideas can still be captured in comparative research, as illustrated below (and as clarified in discussions with Elkin, Stone, and others over several years.)

Still, the distinctive hypothesis from this tradition remains: *business wants growth, so government must encourage growth.*

Political Mobilization: Organized Groups Redirect Policy

Another tradition stresses political mobilization of neighborhood and citizen groups in affecting policy. Social movement work (e.g., Gamson 1990) suggests that citizen alienation from authority can spur mobilization efforts. Urban development efforts sparked resistance, starting in the 1960s when urban renewal funding expanded (Anderson 1964; Mollenkopf 1983). This even led some to posit an "inherent conflict between public bureaucracies and neighborhood based groups" (O'Brien 1975, p. 10). Logan and Molotch (1987) stress competition between groups over exchange value versus use value of urban land as central to local development politics. Although they suggest general bias toward prodevelopment policies, they recognize that opposition can emerge from community-based groups organized around use values, for example, opposing development, gentrification, and residential displacement. Such community-based mobilization has repeatedly been documented as an important urban phenomenon in the past 20 years (e.g., Fisher 1984; Boyte 1980; N. I. Fainstein and S. S. Fainstein 1974; Delgado 1986).

These concerns have spawned two main kinds of studies. First are surveys of individual citizens, often in a single locality, which usually correlate social characteristics of individuals with their support for growth-limiting activities (e.g., Gottdiener and Neiman 1981 and others in Chapter Appendix 2). Second are comparative urban studies that may collect some data on growth-limiting policies (e.g., Donovan and Neiman 1992, unpublished), but still rely primarily on (fairly simple) Census data to measure most variables (thus leaving out most specifics of organized groups or political leaders). However, the cost and complexity of obtaining data on (a) citizens' social background characteristics, (b) group mobilization patterns, (c) local leadership, and (d) local growth policy, and analyzing their many interrelations, has meant that minimal work has been done pursuing specific linkages among these variables. Ostensibly contradictory findings have thus emerged, such as Molotch (1976) or Donovan and Neiman (1992) who find professionals supporting growth limits, whereas Albrecht, Bultena, and Hoiberg (1986) report blue-collar workers are strongly associated with growth control. Such contradictions can often be resolved with more complex and contextualized analysis. We extend past work by including Census-type data, but adding original survey data on organized group activities.

One general hypothesis from this tradition is clear: *Organized groups play a distinct role in fostering citizen concern and leadership responsiveness concerning growth-related policies.*

Political Culture: Whose Rules of the Game?

A common theme in Elazar and Zikmund (1975), Clark and Ferguson (1983), and Webber and Wildavsky (1986) is that political systems vary in their rules of the game. Despite common market forces, local differences still emerge. Why? Elazar and Zikmund stress migration by persons from distinct cultures. In related work, one of us has developed a set of propositions about where and why political cultures change (Clark and Inglehart 1990 and Clark and Lipset 1991, stressing the profound shift in the past two decades away from class politics toward a new political culture [NPC]). A key characteristic of the NPC is its stress on separate issues (such as growth control) rather than joining them into a general New Deal liberal or conservative program. Similarly, the NPC stresses social issues and consumption over economic and fiscal issues, which implies that growth control illustrates a class of issues that should emerge and succeed because citizens increasingly want to enjoy their surroundings, even if they "lose money" by doing so. We here consider factors encouraging this shift at the city level, using growth-limiting movements and policies as typical concerns of the NPC. Persons are more likely to adopt and implement these NPC patterns in their locales if they have had distinct experiences in their social backgrounds—such as growing up in a time and place where others around them encourage foregoing more traditional growth patterns.

Growth-limiting activities are thus a perfect NPC issue. The NPC approach suggests first that *growth-limiting activities are more likely in cities where citizens and leaders are younger, have more education and income, and have less labor force involvement in agriculture and manufacturing and more in services, especially professional and high-tech fields.*

Besides these socioeconomic characteristics, NPC supporters follow distinct rules of the game in civic and political activities. They reject the hierarchy of strong mayors, bureaucrats, parties, and unions. These are seen as rigid actors unresponsive to the newer trends. Such older, hierarchical actors emerged from tough political battles of class politics—of workers against management, rich against poor. *Class politics persists more in cities with disadvantaged and nonwhite residents,* maintained by tough leaders who stress class politics to mobilize support across a wide range of issues. Yet by stressing class politics, hierarchical leaders often lose touch with the new issues—or do less well than professional ecologists, feminists, the elderly, or others who pursue just their issues. Generalist leaders cannot surpass the specialists on their own issues; their claim to legitimacy is their ability to confront "the establishment" and fight on any issue. Where hierarchical leaders are established and confident enough of their support base not to respond well to individual organized groups, then organized groups are more likely to mobilize independently of elected officials. They should find support in cities with citizens who

BOX 4.1 Summary of Hypotheses

Capitalism and Competition

H1: Local governments must encourage development.

H2: It is irrational for local governments to limit growth.

Strong Business, Growth Machines, and Progrowth Regimes

H3: Business wants growth, and government must encourage growth.

Political Mobilization: Organized Groups

H4: Organized groups play a distinct role in fostering citizen concern and leadership responsiveness concerning growth-related policies.

Political Culture

H5: Growth-limiting activities are more likely in cities where citizens and leaders are younger and have more education and income and have less labor force involvement in agriculture and manufacturing and more in services, especially professional and high-tech fields.

H6: Cities with more minority residents are less likely to support growth limits due to their greater economic deprivations and continued attachments to class politics.

H7: Localities with more "hierarchical" leadership are more likely to generation antigrowth movements.

are concerned more with the "new" issues and find "old" leaders insensitive to them.

Consequently: *localities with more "hierarchical" leadership are more likely to generate antigrowth movements.* We have elsewhere termed this the Hierarchy Leveling hypothesis: hierarchies promote social and political movements that seek to level the hierarchy (Clark 1989; see also discussion of the Hierarchy Leveling hypothesis, Chapter 2).

Summary of Hypotheses

The above hypotheses are partially complementary, each addressing somewhat different independent variables. We summarize them in Box 4.1.

Data Sources on Local Development and Growth Control Strategies

We draw on four studies for development strategies and two for political dynamics.

Fiscal Austerity and Urban Innovation (FAUI) Project

The most extensive study of cities in the world to date, the FAUI project includes surveys in 35 countries. The U.S. FAUI survey was conducted by 26 teams, who surveyed mayors, council members, and administrators in all 1,030 cities over 25,000 (Clark 1989 includes the full questionnaires). About 45 percent of local informants responded to mailed questionnaires and telephone follow-ups, mostly in 1983-1984. Mayors and council members were asked a wide variety of questions about their policy preferences, organized group activities, citizen preferences, and staff autonomy. We and others have constructed measures of urban decision making, as summarized in Chapter Appendix 1. The core survey data have been continually merged with fiscal, population, administrative, and other data—as illustrated by our merging of studies for this chapter.

The FAUI survey asked the chief administrative officer to score 33 strategies for their importance. One was "impose controls on new construction to help limit population growth," which is one of our three growth-related policies analyzed below.

Goetz Survey of Economic Development

This 1989 national survey of economic development officials included 15 development tools reported in Goetz (1990) and another item analyzed here for the first time: "Does your jurisdiction have an active antigrowth group?" This was our measure of antigrowth movement activity.

International City Management Association (ICMA84)

The most comprehensive national survey of local development strategies we have found was that of the International City Management Association (ICMA) in 1984. We constructed four measures from it:

BOX 4.2 The Case Studies: Overview of Seattle, Boca Raton, Boulder, and Los Angeles

The four brief cases, which appear boxed through the text, are nationally important. They have been followed closely as trendsetters by many activists and city officials across the United States. The trend they helped set is toward limiting growth, usually toward "managed" growth, which implies acceptance of moderate growth combined with adequate related facilities (such as infrastructure) and preservation of major natural amenities.

Most of these towns fit our political culture hypotheses and national findings in Tables 4.3 to 4.6 in that they have relatively affluent and highly educated residents, often a university that is the major employer (Boca Raton, Boulder), and a high-tech/professional occupational structure. All have striking physical surroundings, which attracted many residents, who have then sought to conserve them. Los Angeles resembled the other locations more in the past than by the 1980s, when its size was overwhelming. It plays a major role at the center of dramatic growth versus antigrowth debates in Southern California. The major growth-limiting groups were located in Los Angeles areas that fit our general characterization: Santa Monica, Pacific Palisades, and Westwood around UCLA, with nearby beaches and striking canyons.

All our case study locations have substantial campus-like, high-tech parks that house international leaders in computer hardware and software development, and other employment activities that link to nearby universities. Examples: the West Los Angeles area was home for Peter Norton Utilities, Quarterdeck-QEMM, and other software developers; Seattle is home of Bill Gates and MicroSoft, the top software company in the world; Boulder has IBM, publishing companies, Celestial Seasonings—the leading U.S. herb tea—and upscale beer breweries; Boca Raton has IBM. And San Francisco, discussed in the text, has its Silicon Valley suburbs. Related financial, legal, accounting, and professional activities have flourished. The firms just listed have also been repeatedly covered in national news media like *Fortune* as defining the style of future U.S. business leadership. Their founders and top executives are portrayed as building their firms from scratch while in their 20s and 30s; as sporting casual dress in laid-back surroundings, but having intellectual brilliance; and as reaching multimillion dollar success by their 40s.

"Lifestyle" is critical in these locations; residents often take bike paths more seriously than freeways: streets and parks are filled with joggers, Rollerbladers, and skiers. "Consumption" and "production" often join in these towns, which are national leaders in casual clothes, sports equipment (Eddie Bauer in Seattle), and sports medicine—with physicians and therapists, special plastics and related medical products.

As an addendum to our boxed case study towns, we note that the eight California cities using the most growth management strategies (San Francisco, Berkeley, Palo Alto, etc. listed in Figure 4.1) offer a profile broadly similar to those discussed here.

- Traditional development strategies
- Minority program encouragement
- Amenities and historic preservation
- Growth control and strict environmental regulation

We find distinct correlates for each type of strategy, supporting the NPC arguments about issue specificity. In this chapter, we mainly consider the last of the four types. Specifics are in Chapter Appendix 1.

California Growth Control Survey (California League of Cities)

The most visible growth-limit discussions are in California, so we merged a survey of 166 California municipalities concerning 10 growth-related strategies, listed in Figure 4.1. Most strategies are discussed in Schiffman (1990).

Other surveys we acquired and analyzed in preliminary manner include the ICMA89 economic development survey, similar to ICMA84 but with fewer strategies, and others by Feiock (1991) and Clarke and Gaile (1991). Each is ideal for certain purposes, but for purposes of this chapter these last surveys simply had low *N*s or fewer items on our central issues.

Model Specification

We specified a recursive path model summarized in Figure 4.2. First we analyzed causes of the Goetz antigrowth movement (X44R). Second we added the antigrowth movement to other variables to estimate their impact on three types of growth control policies of the local government, policies taken from the FAUI, ICMA, and California League studies. Antigrowth "machine" is thus operationalized by the movement plus the three policy measures. Logistic and OLS multiple regression were used to estimate effects.

Results of Data Analysis

We present the main results organized by two sets of dependent variables, following Figure 4.2: first, antigrowth movements and

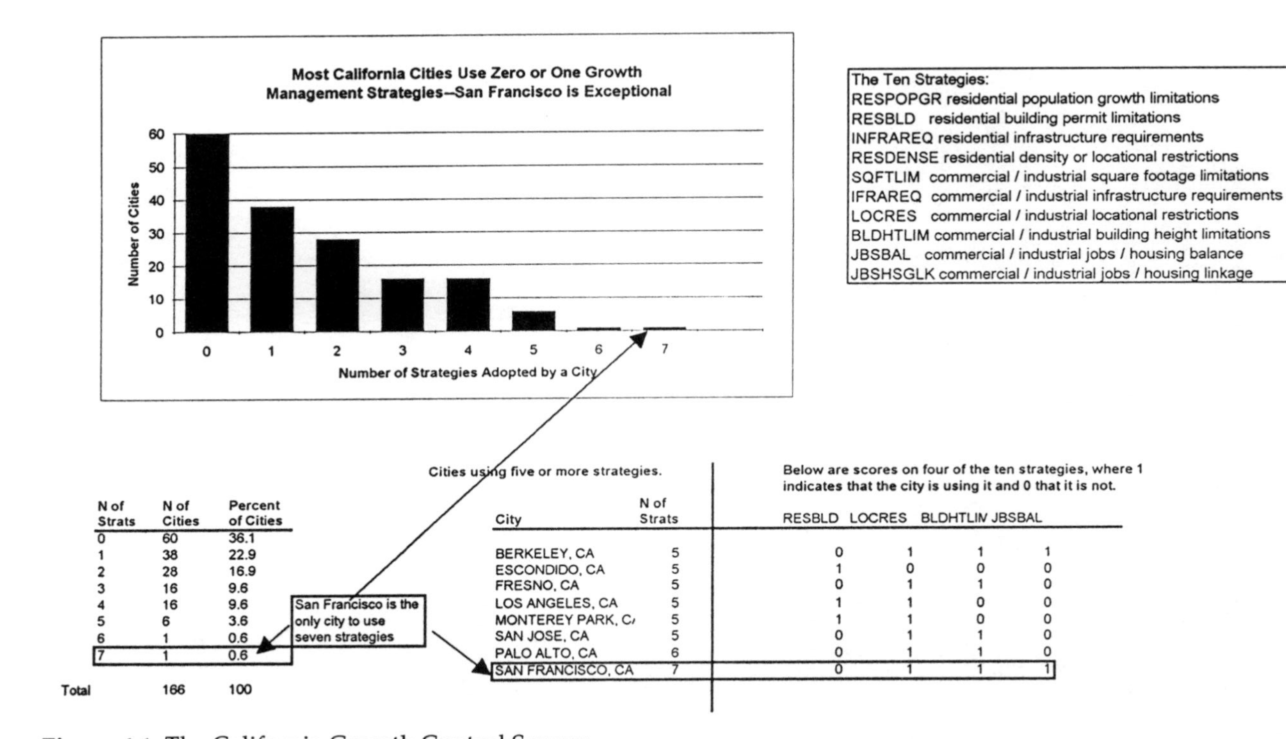

N of Strats	N of Cities	Percent of Cities
0	60	36.1
1	38	22.9
2	28	16.9
3	16	9.6
4	16	9.6
5	6	3.6
6	1	0.6
7	1	0.6
Total	166	100

City	N of Strats	RESBLD	LOCRES	BLDHTLIM	JBSBAL
BERKELEY, CA	5	0	1	1	1
ESCONDIDO, CA	5	1	0	0	0
FRESNO, CA	5	0	1	1	0
LOS ANGELES, CA	5	1	1	0	0
MONTEREY PARK, C/	5	1	1	0	0
SAN JOSE, CA	5	0	1	1	0
PALO ALTO, CA	6	0	1	1	0
SAN FRANCISCO, CA	7	0	1	1	1

Figure 4.1. The California Growth Control Survey
SOURCE: California Growth Control Survey

BOX 4.3 Seattle

In May 1989, Seattle, Washington, voters approved the Citizens' Alternative Plan (CAP), a policy of downtown office space growth control. Like the similar San Francisco measure, the CAP limits the amount of downtown office space for a 10-year period and regulates density and heights in various downtown zones. The CAP was initiated by a coalition of neighborhood groups concerned about the rate of unregulated growth in the city and the emerging problems of traffic congestion, building and population density, and declining quality of life. They stressed the potentially damaging effect of unregulated growth on the character of downtown, the skyline, and the city's infrastructure. City officials earlier attempted to head off the crisis by suggesting that development policy should be made by elected officials. Mayor Charles Royer introduced a growth-limitation ordinance aimed at managing the rate of office space development, and the city council passed a set of interim downtown growth limits. However, voters still approved the ballot initiative, which was more restrictive than the interim limits passed by the council.

The CAP was one of a series of policies urged by community-based activists to mitigate adverse impacts of downtown development and unregulated growth. In 1981, at the urging of housing advocates, the city instituted a replacement policy for housing units demolished downtown. However, this policy was invalidated by the Washington State Supreme Court in 1987. The city developed a modified housing linkage program that offers office developers density bonuses in return for affordable housing development. Citizens also coalesced to delay completion of a freeway linkup between I-5 and I-90 near the city's core.

The CAP limits downtown office space development to 500,000 square feet per year through 1994 and 1 million square feet per year from 1995 to 1999. Density and height limitations limit new developments to 450 feet.

their causes; second, how these and other variables affect growth-limitation policies.

What Drives Antigrowth Movements?

Our measure of organized group activity in the Goetz survey of development officials was "Does your jurisdiction have an anti-growth group?" Responses:

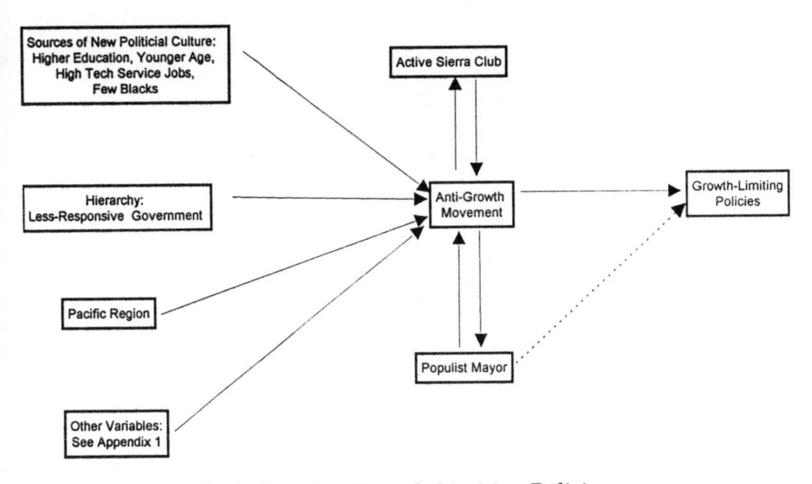

Figure 4.2. Key Paths Driving Growth-Limiting Policies

	Percent	*N*
NO	68.7	123
YES	26.3	47
DK/NA	4.7	9
Total	100.00	179

What kinds of places are more likely to have antigrowth groups? The 47 reporting antigrowth groups include many college towns, such as Austin and Berkeley, but also Baltimore, New York City, and many Florida and California towns; others are locations with nuclear power nearby: Oak Ridge, Tennessee, and Portsmouth, New Hampshire. Figure 4.3 shows higher concentrations in the Northwest, Pacific Coast, and New England.

Can our hypotheses help explain the emergence of these growth control movements?

- How does the finding that 26 percent of our cities have growth movements square with H1 and H2, which suggest that local governments must encourage development and thus antigrowth movements are irrational? This tends to refute H1 and H2.

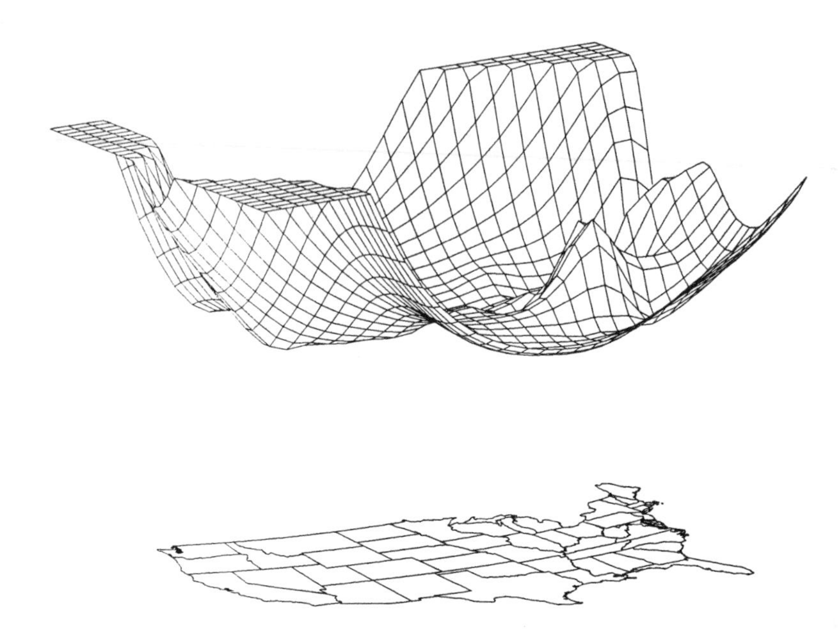

Figure 4.3. The Geography of Antigrowth Movements

- H3 about business power implies that cities with more powerful business leaders should have fewer growth movements. We find no support for H3: business leadership is simply unrelated to having a growth movement. This emerged in the regressions discussed below when business leadership was included as an independent variable. Chapter Appendix 1 summarizes business leadership measures.

- H5 and H6 specify character of places more likely to have antigrowth groups. Several variables from the NPC hypothesis (H5 and H6) are supported in simple correlations (Pearson *r*s). Growth-control movements are more likely in college towns (r = .29), which have more professional and technical persons in the labor force (r = .34), more young persons (r = .18 with age 25-35 and r = .17 with age 35-44), more college graduates (r = .37), higher income (r = .23), more socially and fiscally liberal council members (r =.23, .14), and fewer blacks (r = –.15). This general pattern recurs in regressions, as discussed below.

BOX 4.4 Boca Raton

For years Boca Raton, Florida, attracted elderly and affluent persons from the North, who chose the area for its stunning coastline and lush tropical palm trees and flowers; they did not want to see it diminished. The State located a major new research campus there, Florida Atlantic University. And when hierarchical IBM launced a major research initiative, it created a campus-like facility near the Boca beach, far from IBM headquarters in New York. Result: birth of the IBM Personal Computer, which brought fame to Boca. But so did its early and stringent growth management efforts.

In 1960 Boca Raton was home to 6,000 people. It grew more than 10 times that amount in the next 20 years. The tremendous growth from 1960 to 1980 threatened the access of residents to public beaches and the quality of life. As in other cities profiled here, elected officials were slow to recognize the extent of resident concern. When the City Council failed to control growth, a coalition of older residents, retirees, newcomers, and students from nearby Florida Atlantic University formed to push the growth control agenda.

The coalition gathered 5,000 signatures to call a public referendum on growth. In 1972, voters imposed a growth limit on the City of 40,000 housing units, intended to keep the population at about 105,000 (Meador 1978). City officials implemented the growth limit policy through a range of programs including the public purchase of beachfront recreational land, downzoning, and development moratoria.

• H7 suggests that more "hierarchical" leadership should generate antigrowth movements. Hierarchy we identify here via three measures of nonresponsiveness of political leaders. The FAUI survey asked mayors and council members how often they responded to a list of organized groups and to individual citizens (full text in Chapter Appendix 1). We analyzed these first by summing the responsiveness reported by the mayor and council members to each type of organized group—not just antigrowth movements. Second we summed the mayor and council's reported responsiveness to citizens (MYCCITZM). Third we analyzed a populism item: Q30. "About how often would you estimate that you took a position against the dominant opinion of your constituents?" A consistent pattern emerges across all three: *antigrowth movements are more common in cities where mayors and council members are less responsive* (see Figure 4.4). The

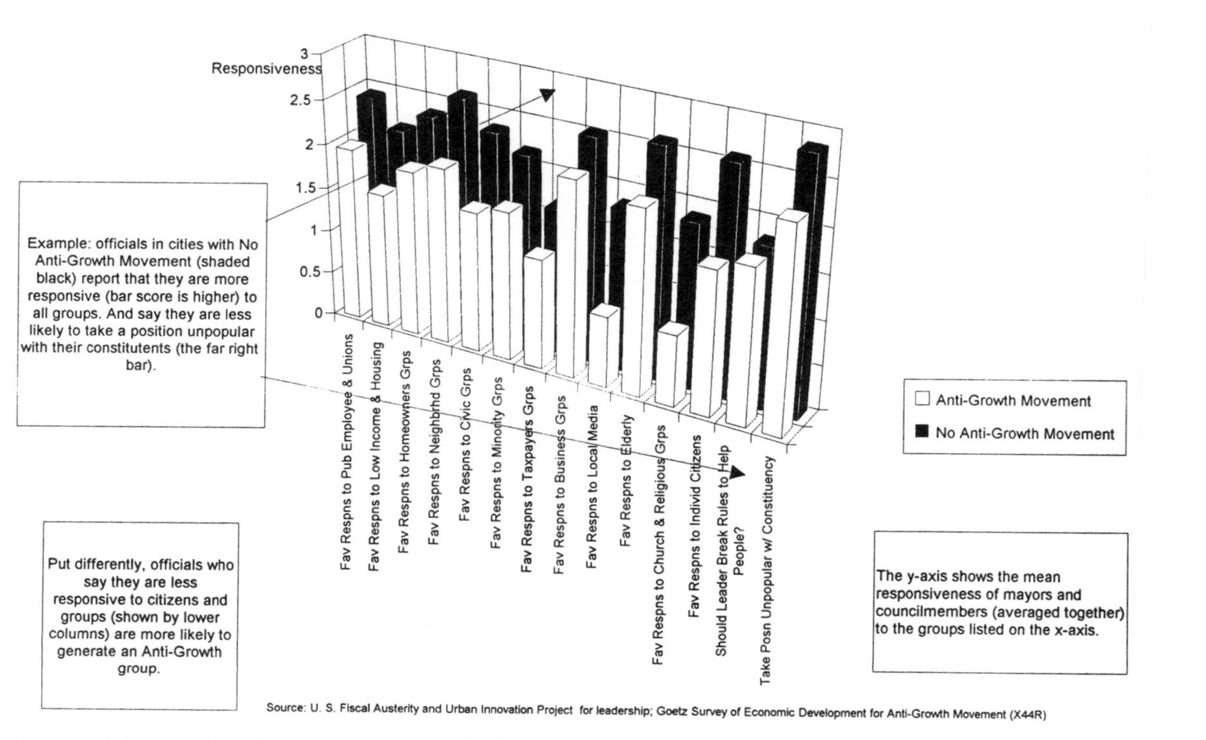

Figure 4.4. Cities With Mayors and Council Members Who Are Unresponsive in General Are More Likely to Have an Antigrowth Movement: Hierarchy Encourages Social Movements

pattern holds for many types of groups—civic and church groups, the media—but especially citizens. (Figure 4.4 lists categories of organized groups.) That these political process items came from mayors and council members rather than antigrowth activists makes the results all the more striking.[4] Regressions below show the same pattern. H7 is thus supported.

We also considered region: cities on the Pacific coast had more antigrowth movements, controlling for other factors. Some past observers note the same regional trend in antigrowth movements (e.g., Schneider 1991) and policies (e.g., Population-Environment Balance, Inc. 1987). Why? Consider alternative interpretations:

- *Physical geography.* Pacific area residents may be encouraged to curtail growth by the natural physical attractiveness of their region. This may be plausible, and not in conflict with other hypotheses, but hardly is sufficient. For example, many underdeveloped countries have phenomenally attractive surroundings that their poor residents do not preserve. Geography must interact with other variables, such as income, to have such sociopolitical effects as generating a growth movement.

- *Recency of migration.* More recent migrants, who were attracted enough by an area to move there, are more likely to seek to preserve its distinctive qualities; and there are more migrants in the West.[5] The rapid and recent growth in the Pacific region is grist for activists who decry "Californication" of the landscape. This led us to test several measures of population growth, but they proved unrelated to our antigrowth movement measures.

- *Laws.* There are more legal provisions for referenda, recall, and other populist inputs to government in the West. But why?

- *Regional political culture.* Residents in the Pacific area may hold different beliefs about politics. We might expect, with Elazar and Zikmund (1975), that the region should differ due to being settled via the distinctive migration patterns of moralistic New England Protestants who moved to much of Northern California, Oregon, and Washington. They often brought different rules of the game about government and civic groups—favoring activist citizens and abstract moral issues—consistent with the Sierra Club and similar groups. To capture

BOX 4.5 Boulder

Boulder, Colorado, has one of the nation's longest histories of growth management. As far back as 1959 a charter amendment established a "blue line" up to which city utility services would be extended. Beyond the blue line, however, utility services were not offered, effectively halting development. Beginning in the 1960s, environmental and antiwar activists grew more active in local political issues, including development controls to preserve the city's natural amenities. In 1967, voters approved a program to fund creation of a greenbelt around the city fringes to preserve its open spaces. In 1971 the local chapter of Zero Population Growth (ZPG) sponsored a ballot initiative to limit the city's population to 100,000. Although the initiative lost, the Boulder City Council adopted a set of growth management policies to reduce the rate of growth. A citizen commission was created to study the area and alternative growth scenarios for the region. In 1973 the commission proposed specific growth control measures. In 1976 Boulder voters approved a ballot measure to create an ordinance limiting the number of residential permits to 450 annually.

The acceptance of environmental concerns and growth management ideology became so complete in Boulder that public debate ended on whether growth should be limited. A 1976 report stated that "left, right, and center Boulder citizens assent to [growth control] ideas (Penne 1976, p. 11)." The real estate industry and the Chamber of Commerce agreed with environmental groups that controlled growth was the best path for the city. The debates in Boulder in the 1980s and early 1990s concerned specific methods to control growth.

the specific role of the Sierra Club, we obtained from it the membership and number of letter writers for each of our localities. The two Sierra Club measures correlate r = .3 with the Goetz antigrowth movement measure.

The three sets of variables just discussed—NPC socioeconomic characteristics, hierarchy, and Pacific region—hold strong in several multiple regressions. The causal pattern they generate is summarized in Figure 4.2. To consider all explanatory variables in a regression-type analysis causes problems with limited cases. We thus chose one key variable from each of the three clusters and added selected other variables in different combinations. Occasional additional variables were significant, but these three basic clusters remain important in several alternative models. Table 4.1 reports detail: antigrowth groups are explained by percent college educated (PV871), less responsive mayors and council members (MYCCITZM), and Pacific region (PAC). Many other variables are weak (population size $r = .09$) or insignificant (population growth from 1980-1986, percent nonwhite, Elazar's moralistic political culture (PKMORAL), an Atkinson (1970) index of inequality of educational attainment among citizens (ATKTOT1), city manager government form, range of functions performed by the city government (results not shown). We find no relation in correlations or regressions between business leadership and antigrowth movements.

Conclusion: the presence of antigrowth groups is poorly explained by H1 or H2 about capitalism or business leadership. Our boxed case studies provide counterexamples, illustrating more specific alternative dynamics. These are not purely idiosyncratic exceptions, but leaders of a substantial national movement—our survey finds antigrowth movements in 26 percent of U.S. cities. H3 on business leadership and growth machines is contradicted most forcefully by the correlations in Table 4.4 and subsequent regressions: they show zero impact of business on antigrowth activities. We do find support for H5, H6, and H7, the political cultural hypotheses. Hierarchical leadership engenders antigrowth movement, especially when citizens are young, educated, wealthy, nonminority, high-tech professionals and service workers.

This produces an antigrowth machine.

What Drives Growth Policies of City Governments?

We now ask what drives growth-limiting policies and move one causal step further in our path diagram (Figure 4.2). The hypotheses discussed above apply here, as does H4, which was not applicable in the last section: *H4: Organized groups play a distinct role in fostering citizen concern and leadership responsiveness concerning growth-related policies.* To measure organized groups in this section we use mainly the Goetz growth item, the dependent variable in the last section (X44R in Table 4.1). Is H4 supported? Yes, indeed. By contrast, variables testing most other hypotheses remain weak or insignificant in regressions, although some are significant in correlations. These are the key findings of this section, which we now elaborate.

TABLE 4.1 Sources of Antigrowth Movements: OLS and Logistic Regression Estimates

Dependent Variable: X44R
Antigrowth Movement
OLS Estimation

Multiple *R*	.65
R Square	.42
Adjusted *R* Square	.37
Standard Error	.36

Independent Variable	*B*	*Beta*	*Sig T*
Nonhierarchical Council (CITRESP + CITRESC/2), MYCCITZM	–.33	–0.46	.00
Percent w/Ed 4 Yrs College, PV871	1.64	.37	.01
Pacific Coast, PAC	.11	.29	.04
(Constant)	–.09		

Logistic Estimation

	Chi-Square	*DF*	*Signif.*
(-2Log) Likelihood	24.98	33.00	.84
Model Chi-Square	13.65	3.00	.00
Improvement	13.65	3.00	.00
Goodness of Fit	34.49	33.00	.40

Classification for X44R, Percent Correct Overall 86.49%

	B	*SE*	*Wald*	*DF*	*Sig*	*R*
Nonhierarchical Council (CITRESP + CITRESC/2), MYCCITZM	–1.91	.93	4.22	1.00	.04	–.24
Percent w/Ed 4Yrs College, PV871	10.96	5.76	3.62	1.00	.06	.20
Pacific Coast, PAC	.69	.54	1.67	1.00	.20	.00
(Constant)	–4.26	1.44	8.73	1.00	.00	

SOURCE: U.S. Fiscal Austerity and Urban Innovation Project and Goetz Survey of Economic Development.
NOTE: See Figure 4.4 on the Hierarchical Council measure; the measure here is citizen responsiveness, which we thus label "Non"hierarchy. Given the low *N* in this table we reestimated several alternative models, adding and deleting each of these three plus selected other variables reported in the text. OLS results are shown for simpler communication, but so are logistic regression results because the dependent variable is dichotomous. In the logistic model here Pacific region falls below significance; it is significant when MYCCITZM is omitted, even though PAC is uncorrelated with MYCCITZM (their Pearson *r* = .01). See Chapter Appendix 1 for more detail. Given that OLS regression and significance tests assume a normally distributed dependent variable, the OLS results should be interpreted cautiously.

TABLE 4.2 Correlations Between Antigrowth Movement and Growth Control Policies

	ICMA84 Growth Control Strategies (GRCONIX)	*FAUI Growth Control Strategies (V262R)*
Antigrowth Movements (X44R)	.24** (N=76)	.19* (N=100)

SOURCE: Goetz Survey of Economic Development (X44R), ICMA 1984 survey (GRCONIX), and U.S. Fiscal Austerity and Urban Innovation Project (V262R).

	California Growth Control Survey Items			
	RESPOPGR	*RESBLD*	*INFRAREQ*	
Antigrowth Movements (X44R)	.26*	.43**	0.18	
	RESDENSE	*SQFTLIM*	*IFRAREQ*	*LOCRES*
Antigrowth Movements (X44R)	0.06	0.18	.25*	.43**
	JBSHGLK	*BLDHTLIM*	*JBSBAL*	
Antigrowth Movements (X44R)	0.09	.38**	.32**	

SOURCE: California Growth Control Survey.
NOTE: * = significant at .10 level; ** = sig. at .05 level; *** = sig at .01 level. *N* is 28 for the *r*s of all California Growth Control Survey items. Definitions of the Ten California Growth Strategies: RESPOPGR residential population growth limitations; RESBLD residential building permit limitations; INFRAREQ residential infrastructure requirements; RESDENSE residential density or locational restrictions; SQFTLIM commercial/industrial square footage limitations; IFRAREQ commercial/industrial infrastructure requirements; LOCRES commercial/industrial locational restrictions; BLDHTLIM commercial/industrial building height limitations; JBSBAL commercial/industrial jobs/housing balance; JBSHSGLK commercial/industrial jobs/ housing linkage.

Our three sources of growth-limiting policies—FAUI, ICMA, and California League of Cities—each tap a slightly different dimension. The FAUI item asks officials if they "impose controls on new construction to limit population growth" as a strategy for managing fiscal problems. The ICMA items tap procedural matters, specifically whether the city requires environmental impact statements, has altered any strict environmental regulations, and so on (see Chapter Appendix 1 for the full items). The California League of Cities survey includes a wider range of strategies—10—but only for California cities. We thus report them individually in Table 4.2 and in an index summing the 10, CALGROW, in Table 4.3.

Despite the range of items and their specifics, one powerful and consistent result is clear: *antigrowth groups cause growth control policies.*[6]

TABLE 4.3 Correlations of Three Growth-Limiting Strategies With Other Variables

	ICMA84 Growth Policies GRCONIX	*FAUI Control Strategy V262R*	*California Growth Index CALGROW*
Pct. Population w/4 years College PV871	.07*	.12	.21***
Log per capita income LIPCINCT	.24***	.14*	.08
New Political Culture Index NPCIX	.16**	.29**	.09
Prof/tech employees per 1000 residents PROFTCEM	.04	.14*	.21***
Membership in Sierra Club AVGMEMSH	.36***	.13	.16**
Pct. Below Poverty ITEM40	–.17***	–.09	.05

SOURCE: ICMA 1984 survey (GRCONIX), U.S. Fiscal Austerity and Urban Innovation Project (V262R and other variables), California Growth Control Survey (CALGROW).
NOTE: These are Pearson *r*s. * = significant at .10 level; ** = sig. at .05 level; *** = sig. at .01 level. NPCIX sums five components as defined in Clark and Ferguson (1983): fiscal conservatism, social liberalism, nonresponsiveness to organized groups, responsiveness to citizens, and support for public goods-type policy outputs.

Table 4.2 shows clearly positive correlations between antigrowth group and (a) the ICMA Growth Control Policies, (b) the FAUI Growth Control Strategy, and (c) 6 of the 10 California Growth Control Survey items.

H6 and H7 concern organized groups, citizens, and political culture as driving policy. Let us look more closely at their specific effects on growth-limiting policies. Past studies that seek to explain growth control policies have seldom used an organized group measure like our antigrowth movement. Population and economic variables are more common (see Chapter Appendix 2). We thus examined many variables used in past work in correlations and multiple regressions, along with the antigrowth movement measure, to be sure that its effects were not spurious. Several variables correlated significantly with the three antigrowth policies (Table 4.3): at least one of the three antigrowth policies was more likely in cities with more college graduates, higher income, more professional and technical employees, higher Sierra Club membership, fewer black residents, and higher scores on the New Fiscal Populism index, and the results hold up in regressions.

TABLE 4.4 Business Leadership Is Unrelated to Growth-Limiting Policies

		Growth-Limiting Policies		
		ICMA84	*California*	*FAUI*
		Growth	*Growth*	*Growth*
	Antigrowth	*Control*	*Control*	*Control*
Four Business	*Movement*	*Policies*	*Index*	*Strat.*
Leadership Measures	*X44R*	*GRCONIX*	*CALGROW*	*V262R*
V97R	–.20	.03	–.02	–.14
Responsiveness to Business	[79]	[193]	[56]	[307]
(closed end)	P=.04	P=.33	P=.45	P = .01
BUZDIX1	.05	.07	–.05	.06
Frequency of Mention of	[66]	[174]	[54]	[244]
Business (open end)	P = .34	P = .19	P = .35	P = .16
BUZOT	–.09	.05	.01	.01
Relative Frequency of	[66]	[174]	[54]	[244]
Mention of Business (open end)	P = .24	P = .24	P = .47	P = .43
MYCV97M	–.13	–.06	–.62	–.16
Mayor and Council Average	[39]	[103]	[3]	[186]
Responsiveness to Business	P = .21	P = .29	P = .29	P = .02

SOURCE: Goetz Survey of Economic Development (X44R), ICMA 1984 survey (GRCONIX), and U.S. Fiscal Austerity and Urban Innovation Project (V262R and business leadership measures), California Growth Control Survey (CALGROW).
NOTE: The table shows for each matrix entry at top the Pearson *r*, middle in brackets the [*N*], and bottom *P* = the significance level. The significant *r*s here fall to insignificance in the regressions. Two business leadership measures used open-ended items, and two closed-ended items, discussed further in Chapter Appendix 2. The closed-ended items were V97R and MYCV97M. The two open-ended measures included first the number of mentions of business, BUZDIXI; second was BUZOT=BUZDIXI divided by the frequency of mention of all other participants. The business leadership measures were also unrelated to antigrowth movements in most correlations, and in all regressions specified by forcing the business measures to be included as explanatory variables for the three growth control policies listed in this table.

The regressions testing sources of growth-limitation policies show few strong or consistent results except for X44R, antigrowth movements (Table 4.5). The modest effects of other variables we find are consistent with past work on explaining local development policies (e.g., Feiock 1991; Schmenner 1982; and G. P. Green and Fleischman 1991, which analyzes the same ICMA84 data we use).

Strong in most regression models explaining the FAUI growth control item is the New Fiscal Populist index, constructed following Clark and Ferguson (1983) (Table 4.6, but it falls below significance when X44R, antigrowth movements, is added; see Table 4.5). This illustrates two alternative causal paths to growth control: by organized

BOX 4.6 Los Angeles

In Los Angeles the growth control movement emerged from activities of homeowner groups aimed at slowing downtown development and preserving neighborhood amenities. In 1985, homeowner groups enjoined the city from permitting downtown development that exceeded specifications of the city's general plan. The following year the same groups mounted a growth control initiative, Proposition U, to reduce commercial density by half and require the city to adopt a growth management plan. Proposition U passed despite the opposition of Mayor Bradley and Council leader Pat Russell. In 1987, Russell, Bradley's closest ally on the city council, lost her seat to a growth control advocate. Expecting an electoral challenge from another slow-growth council member, Bradley became a vocal advocate of community-based planning efforts.

The slow-growth movement in Los Angeles was fueled by the efforts of homeowners' associations in the middle- and upper-middle-class San Fernando Valley and the West Side. The movement's targets were increasing density, the threat of declining property values, and maintenance of neighborhood amenities. Hyperinflation of the housing market in Los Angeles during the mid-1980s significantly increased the financial stakes for home and property owners. This preservationist orientation explains "how conservative homeowners in the age of Reagan came to advocate a structural reform implying massive regulation of one of the most sacred marketplaces (Davis 1990, p. 188)."

A second impetus for growth management in Los Angeles has been environmental concern over waste management; sewage treatment; and the quality of air, beaches, and ocean. In 1988, the city was forced to adopt interim building permit controls for projects using the city's sewage system because the system, built in the 1920s and upgraded in 1954, was rapidly reaching its capacity. In 1987, the system had already malfunctioned, dumping millions of gallons of waste into Santa Monica Bay. Environmentalists, concerned about water quality and saving the bay, joined in the chorus of voices advocating stricter building controls. By 1990, the city had imposed a permanent ordinance controlling the rate of approval of building permits.

group activity or political victory. Protest by organized groups often declines when citizens sense that they have a sympathetic leader in city hall. The converse of this is that nonresponsive leaders generate antigrowth protest (as shown in Table 4.5 and our Figure 4.4 hierarchy results). This is consistent with similar findings for other social movements, especially civil rights groups, which have been fre-

TABLE 4.5 Modeling Causes of Three Measures of Growth-Limiting Policies

Dependent Variables	*ICMA84* GRCONIX	*California League* CALGROW	*FAUI* V262R
Multiple *R*	.52	.51	.23
R Square	.27	.26	.05
Adjusted *R* Square	.21	.09	–.03
Independent Variables	*B*	*B*	*B*
Pacific Coast PAC	.30**	—	–.02
Prof/TechEmp/1,000emp PROFTCEM	.00	.00	.00
Average Group Activity GRPACT	.00	.00	.00
New Fiscal Populist Political Culture Index for Mayor, NPCIX	–.15	.02	.08
Per Capita Income (log) LIPCINCT	.93	–.04	–.03
Antigrowth Movement X44R	.36	.17**	.37
(Constant)	–9.16	.43	1.61

SOURCE: Goetz Survey of Economic Development (X44R), ICMA 1984 survey (GRCONIX), and U.S. Fiscal Austerity and Urban Innovation Project (V262R and other measures), California Growth Control Survey (CALGROW).
NOTE: See respecification of the model for V262R in Table 4.6. Pacific is omitted for CALGROWTH model as only California cities are included. With slightly fewer independent variables, X44R rises in significance toward the Table 4.2 results. * = significant at .10 level; ** = sig. at .05 level; *** = sig. at .01 level.

quently studied: they often decline in fervor after a city hall victory, especially election of a first black mayor in a city where the previous administration was considered racist (cf. Clark and Ferguson 1983; Browning, Marshall, and Tabb 1984; Bobo and Gilliam 1990).

Our interpretation thus differs from the more narrow social mobilization/social movement literature (e.g., Gamson 1990) or classic group theory (e.g., Truman 1951; Lowi 1979), both of which stress organized groups—as in H4. We stress instead that organized groups are one means to affect policy, but winning in city hall is a second—as occurred in several of our case studies, especially San Francisco. Mayors inside city hall can be far more effective than protesting groups outside. But mayors often win after growth issues have been added to the local agenda by antigrowth groups—as repeatedly occurred in our case studies of Boca Raton and Boulder, for example (see Boxes 4.2, 4.4, and 4.5). Similar examples of antigrowth mayors winning office in the 1970s and 1980s include Irvine, California, and Bennington, Vermont (cf. Rosdil 1991).

TABLE 4.6 When Antigrowth Groups Are Omitted From the Model, the New Fiscal Populist Mayor Drives Growth Control

Dependent Variable: Control New Construction, FAUI Q8.33, V262R

Multiple *R*	.14			
R Square	.02			
Adjusted *R* Square	.00			
Independent Variables	*B*	*Beta*	*T*	*Significance*
New Fiscal Populist Political Culture Index for Mayor, NPCIX	0.10	.14	2.27	.02**
Pacific Coast, PAC	0.02	.03	0.56	.57
Prof, SpecTechEmp/1,000emp1980, PROFTCEM	0.00	.00	0.06	.96
Average Group Activity, GRPACT	0.00	.00	0.02	.98
Per Capita Income (log), LIPCINCT	−0.03	−0.1	−0.11	.91
Constant	1.62	2.08	0.78	.44

SOURCE: U.S. Fiscal Austerity and Urban Innovation Project.
NOTE: When this regression is reestimated deleting one variable at a time, the New Fiscal Populist mayor result remains consistent and the adjusted R^2 rises. The other variables are included here to show that they do not suppress the NFP mayor. * = significant at .10 level; ** = sig. at .05 level; *** = sig. at .01 level.

A Dramatic National Case: San Francisco

San Francisco is a critical case for several reasons. It has implemented probably the most forceful antigrowth policies of any major U.S. city. Its specifics challenge most theories of growth and seem to match remarkably well the national pattern we find in our regressions. One of us (Goetz) worked on growth-related issues as staff for the City of San Francisco for 3 years. San Francisco has recently been thoughtfully studied by DeLeon (1992), but on growth control we depart from his interpretation.

San Francisco illustrates the two alternative modes for adopting growth control policy that we have identified: via an antigrowth movement or via elected officials. Despite the city's image as a progressive "left coast" city (DeLeon 1992), from 1965 to 1992 the mayor's office was occupied by progrowth administrations for all but 7 years (3 by Moscone and 4 by Agnos). Mayor after mayor who was nonresponsive to antigrowth groups motivated the groups to mobilize and continue to press their case in a social movement, antiestablishment manner. For 25 years San Francisco thus had one of the more active and successful local growth control movements in the nation. And it pursued essentially an oppositional position for most of its existence. The movement's most notable achievements—introduction of an office-housing linkage program and passage of Proposition M, which severely restricted the amount of office space construction in the city—occurred while (generally) progrowth Mayor Dianne Feinstein was in office.

The antigrowth movement emerged with opposition to freeway construction in 1965, followed by a struggle to stop the Transamerica pyramid, built in 1972 (Hartman 1984; Elberling unpublished). These two organizing efforts were ad hoc, but contributed to a growing movement and awareness among neighborhood groups and "urban environmentalists" that the city's development policy threatened to alter its landscape forever. Early on in the mobilization process, downtown high-rise development was singled out as the most prominent development problem facing the city. Between 1965 and 1979 the city added about 8 million square feet of office space every 5 years. The antigrowth coalition warned against the "Manhattanization" of San Francisco and argued for strict controls on the rate of high-rise construction and for height limits on downtown buildings. The movement grew slowly in the initial years and suffered several setbacks, losing three separate antigrowth initiatives in 1971, 1972, and 1979. Antigrowth activists were more successful in opposing specific projects, such as the waterfront freeway and the U.S. Steel building, and in using the 1972 California Environmental Quality Act (CEQA) to oppose downtown development (Hartman 1984). Antigrowth attorneys repeatedly challenged the impact analyses of specific buildings, stopped development of some projects, and extracted compromises from other developers with the threat, or reality, of legal action. They argued more generally that the environmental review process ignored the cumulative impact of high-rise development in the city, and this position was upheld by the courts. These are all examples of confrontation tactics, opposing city leadership. Nonresponsiveness by city hall seemed to spark further movement activity on antigrowth issues, just as it did on numerous other social issues that were actively pursued in analogous antiestablishment social movements in these years. But for two brief periods the movements helped place a champion in city hall.

The 3-year administration of George Moscone (1975 to 1978) illustrates our second motor driving growth policy: the committed and successful political leader. As in many cities with such leaders (see our boxed examples), Mayor Moscone developed his growth-limiting policies by building on social movement examples. But he made them his own policies. This made it unnecessary for antigrowth movements to work against and outside city hall as in the past. Moscone is credited with consolidating the antigrowth movement and coalescing neighborhood-based groups, advocates for the poor, and environmentally oriented antigrowth movements into a broader force for controlled growth (DeLeon 1992; Elberling unpublished). This coalition of forces led the movement to stress the social and environmental impacts of continued high-rise development and to offer solutions mitigating these impacts. Their chief concern was the escalating cost of housing and continued displacement of low-income residents as downtown development forced a tighter and more costly housing market. Their proposed solution, the Office-Housing Production Program (OHPP) was the first office-housing linkage program adopted in a major city in the United States. Mayor Art Agnos (1987-1991) continued similar concerns.

Despite the nurturing effect of the Moscone administration, the antigrowth movement in San Francisco emerged primarily as an oppositional group and had to conduct most of its business as an outsider attempting to influence government policy. That the movement was forced to attempt four initiative campaigns is clear testimony that antigrowth was not supported by top city officials. Further, use of CEQA as a legal means to slow development was refined in the early Feinstein years as an alternative to policy-related growth controls. The office-housing linkage program was similarly proposed in the Feinstein years by community advocates attempting to slow downtown development. Though the program provided millions of dollars for affordable housing that mayors could subsequently spend on "their" policies, it was initially proposed to slow downtown development—by antigrowth activists dissatisfied with the city's continuing development (Goetz Survey 1989).

This combination of a strong antigrowth movement for many years, combined with two supportive mayors for 7 years, led San Francisco to adopt the most stringent policies not only among major U.S. cities, but even compared to smaller, more homogeneous and affluent political oases such as Berkeley, Palo Alto, Santa Monica, or Santa Barbara—as Figure 4.1 indicates.

Discussion and Conclusions: Determinants of Growth Limitations in U.S. Cities

Let us reassess the hypotheses with which we began in light of our findings and related work by others.

Capitalism and Competition

The "structural" hypotheses H1 and H2 suggest that local governments must encourage development and that it is "irrational" to limit growth. Note that some local officials themselves find it easier to invoke "the feds" or "the market" to help say no, as elaborated in Appleton and Clark (1989). The rhetoric of market constraints sometimes sounds deterministic. We argue instead that competitive market forces are indisputably important, but not deterministic. Even for individual firms, market forces are far from deterministic; there are many potential market niches and continual innovation by firms trying out new products and approaches. Local loyalties and nonmarket preferences of managers also affect firm behavior. And market forces, as Keynes reminded us, operate in the "long run"; much can happen during shorter runs. So even among cities seeking growth, the range and diversity of strategies used is considerable. If there is looseness in market constraints on private firms, these constraints are far, far looser for local governments, as a wide variety of firms may locate in a single city and local leaders have many choices as to what types of firms to seek to attract and how. This looseness is simply and powerfully documented by our surveys' descriptive results, which make one consistent point: cities vary in policy profiles. Some support growth; a few are antigrowth; most are mixed. Some have growth control movements; others do not.

The same looseness recurs in our causal analyses and research by others on determinants of urban policy: statistical relations are only moderate; the variance explained in most regression-type models may be 10 or perhaps 30 percent, not more. The policy implication is clear: there is much room for local "political choices" (the fortuitous term of Wong 1989 to reconceptualize Peterson's strong argument). The capitalism and competition hypotheses, H1 and H2, are overstated and, taken literally, are simply incorrect.

Strong Business/Growth Machine

Note that H3—Business wants growth, and government must encourage growth—builds on two implicit assumptions: that business wants (a) growth and (b) government intervention to encourage growth. Both assumptions are problematic. Surely some business leaders support growth, such as in Atlanta, Dallas, and other growth-oriented towns. But how consistently does business support growth? Probably most developers do. But in other firms, commitment to growth is less than obvious. Even Logan and Molotch (1987) argue that growth machine coalitions are subject to political strains and can break down over specific differences in growth strategy preferences. One fault line they identify (following the community power tradition, e.g., Aiken and Mott 1970) is between locally based (usually smaller) businesses and absentee-owned branch facilities of nonlocal businesses. The inequitable nature of many government development incentives also creates conflict among businesses (for example, see Bennett 1986). Further, there are many loose links in the logical chain from business leaders' preferences to public policy to land values. These linkages may operate, but far less forcefully than some accounts suggest (e.g., Elkin 1987).

The assumption that business leaders want growth implies that business leaders conflict with others, especially citizens. The critical question here is how different are business leaders' preferences? This question is casually dismissed in most theoretical works and even most empirical research with the assumption that business leaders want something different from other local participants. Yet, searching for direct evidence either way, we find remarkably little. The null hypothesis is that business leaders are no different from citizens or organized groups in the same locale. "Leaders" often follow as much as lead, especially on controversial public policies like supporting or opposing growth.

Unfortunately, our national growth control surveys offer little direct evidence on these two key assumptions. Consider therefore a few findings from others. Baldassare and Protash (1982) compared business and citizen policy preferences on population growth and found a huge .86 correlation across 97 California communities. Many surveys report that business leaders stress conflict avoidance in their civic leadership roles (e.g., Aiken and Mott 1970). Williams and Zimmermann (1981) show that business policy preferences on antipoverty programs varied enormously across cities. But although they found business leaders' preferences almost unpredictable nationally, local business preferences were often explained by preferences of other local participants. This is consistent with the view that business leaders *adopt* dominant local preferences, especially on controversial public issues. Schumaker (1991) studied nine issues in Lawrence, Kansas, and found few differences between business leaders and other participants. He sought to analyze business and social leaders separately, but found (another huge) .77 correlation between their issue positions (p. 44). On the key issue of growth, he again found business leaders close to other leaders. Perhaps even more compelling, he directly surveyed citizens (unlike most urban elite theorists). Results? The opposite of what business elite theorists imply: *citizens* in Lawrence generally supported growth. Analyzing the ICMA1989 survey of economic development strategies, DeSantis (1991, Table 3.5) found that the importance of business is related to just one of nine types of economic development strategy in multiple regression analyses. State-level work is mixed; some studies suggest impacts of business on policy (e.g., Hicks, Friedland, and Johnson 1978), but Saiz (1990, p. 21) found an impact by labor but not by business on state economic development strategies. See also Gottdiener and Neiman (1981) in Chapter 2 Appendix.

Our key test of the growth machine hypothesis is to ask if towns in which business leaders are more powerful have more active prodevelopment programs or more actual growth. Result: no, when we entered several Business Leadership measures in regressions as explanatory variables for antigrowth movements and our three growth-limiting policies, business leadership was consistently insignificant.[7] Business leadership was also insignificant in regressions explaining "traditional" development programs (which we analyzed similarly to the three growth-limiting policies; see Appendix). Schneider analogously found no relation between the importance of the Chamber of Commerce and amount of growth (1991, p. 10, 1992).

Political Mobilization

Logan and Molotch (1987) suggest that conflict over use and exchange values is central in urban development. DiGaetano (1990) argues that community-based advocates represent the primary alternative to conventional incentive-based economic development (see also Feagin and Smith 1987). Departing from this group mobilization hypothesis, Schneider and Teske (1992) support a second view: that antigrowth political entrepreneurs are an important precursor to antigrowth policy adoption. They suggest that the concept of a political entrepreneur is essential to a more developed theory of citizen mobilization, especially in the antigrowth arena.

We differ from most such past researchers in having data available on each of these factors for many cities. Our analyses permit us to transcend these two orientations. Either organized groups qua social movements or elected officials qua political entrepreneurs can advocate and implement growth-limiting policies. Indeed, we find both groups and leaders important. But how do we conceptually integrate these two explanations? Our contextual relativism integrates the two competing variables in a single more general explanation; shifts in context change the importance of the driving factors. Antigrowth policy patterns differ in two contexts. In the first, citizens mobilize against the nonresponsiveness of local officials; groups then significantly affect policy outcomes. In other contexts, elected officials come to office supporting a political culture of environmentalism and social liberalism. One causal step back, of course, they often reached office and chose these growth-limiting policies with encouragement from active organized groups—as in Boca Raton, Boulder, and San Francisco (see above). In such a context, the elected officials become the policy entrepreneurs, and organized groups recede; groups become "analytically invisible," that is, exert no statistically significant independent effect on policy.

A Short Conclusion

Growth-limiting movements are a strategic research site for recent general theories of urban decision making, as most theories assume widespread support for growth. This assumption is contradicted by our finding that 26 percent of 179 surveyed cities had antigrowth movements and many more adopted growth-limiting policies.

Analyzing sources of these antigrowth movements across U.S. cities, we find little support for the capitalism and competition hypotheses of Peterson et al. or for the strong business, growth machines, and progrowth regime hypotheses of Molotch, Elkin, and Stone. We do find support for the political culture hypothesis concerning the socioeconomic bases of growth-limiting activities, and for the Hierarchy Leveling proposition: hierarchy promotes social movements that seek to reduce hierarchy.

Shifting to growth-limiting policies adopted by local governments, we find that the political mobilization-organized groups hypothesis is the most consistently supported: cities with more active organized groups opposing growth adopt more growth-limiting policies. Political entrepreneurs may supplement these antigrowth movements if they are elected as mayors or city council members. When antigrowth thus becomes the "establishment," the antigrowth movement may decline, but growth-limiting policies nevertheless continue in a more "institutionalized" manner.

Appendix 1
Survey Items, Indexes, and Analyses Not Reported in Text

New variables and indexes were constructed from several complex data sets. The four basic surveys are summarized above. The FAUI survey of all U.S. cities over 25,000 in population included mayors, council members, and chief administrative officers. Mayors and council members were given a series of items that asked about activity and responsiveness to various local participants. The activity item is Q8 (listed at the end of Chapter 2) generating GRPACT. Q9: "Please indicate how often the city government responded favorably to the spending preferences of the participant in the last three years. Circle one of the six answers for each of the types of participants. The city has responded favorably 1 Almost never 2 Less than half the time 3 About half the time 4 More than half the time 5 Almost all the time DK Don't know/Not applicable." The list included 21 participants, including neighborhood groups and others listed in Figure 4.4. The 11 participants shown there (from "Public Employees and Unions" through "Church and Religious Groups") were summed to generate GRPRESC, the general responsiveness index. This is one measure of (non)hierarchy. The individual items from this series for "Businessmen and Business-Oriented Groups or Organizations (e.g., Chamber of Commerce)" and "Individual Citizens" were used to measure business importance and citizen responsiveness respectively.

Two additional measures of business importance were constructed, starting from two open-ended FAUI items: "Q10. The types of participants listed above are frequently active in city government affairs. Could you list the five types of participants which are most influential in decisions affecting city fiscal matters in the past three years? (Feel free to mention participants not on the above list.) [Five blank lines are printed to fill in]." And "Q11. What are the most influential participants in city government in general (not just in fiscal matters)? ____A. Same five as in Q10. ____B. If any are different from Q10, please list all five [Five blank lines printed to fill in]."

Because 73 percent of the mayors answered A. to Q11, we analyzed mainly Q10. From Q10 we computed three kinds of measures of importance for business (as well as for all other groups mentioned). One business importance index was simply the number of mentions of business, BUZDIX1. A second was BUZOT = BUZDIX1 divided by the frequency of mention of all other participants. A third was the average of BUZDIX1 for the mayor and council combined (MYCV97M). Business importance measures were uncorrelated with antigrowth movement (X44R) in most correlations (Table 4.4) and all regressions, e.g., when the business measures were forced to be included in regressions explaining antigrowth movement and the three growth control policies (Tables 4.5 and 4.6).

The ICMA Index of Growth Control Policies was computed as

$$grconix = (\text{enimpctr}*2) - (\text{zoningr} + \text{ennregr})$$

where

enimpctr = "Q7. Has your city ever required developers to prepare an environmental impact statement for their proposed projects? Yes___ No___."

zoningr = "Q6. Has your city undertaken any efforts to reform building or zoning regulations in order to facilitate economic development? Yes___ No___.

If 'YES,' what reforms have been undertaken? (Check all applicable)

a. Consolidated 'one stop' permit issuance

b. Building inspection restructuring

c. Creation of an ombudsman position to facilitate problem resolution

d. Modification of zoning process

e. Administrative negotiation between city and developers on regulators requirements

f. Other (specify)________________________________."

ennregr = "Q8. Has your city altered any of its environmental regulations to facilitate economic development? Yes___ No___.

If 'YES,' how have environmental regulations been affected? (Check all applicable)

a. Relaxed environmental impact statements

b. Eliminated environmental impact requirements

c. Relaxed standards for upgrading infrastructure related to the development (e.g., size and construction of streets and sewers)

d. Allowed for administrative negotiation between city and developers on environmental impact regulations

e. Other (specify)_______________________________."

The California League of Cities conducted a survey of California cities from which we constructed the California Growth Control Index:

$$calgrow = (\text{RESPOPGR} + \text{RESBLD} + \text{INFRAREQ} + \text{RESDENSE} + \text{SQFTLIM} + \text{IFRAREQ} + \text{LOCRES} + \text{BLDHTLIM} + \text{JBSBAL} + \text{JBSHSGLK})/10.$$

The 10 components are defined in Figure 4.1. The 10 fall into four factors in a principal components analysis. However, the main result is that most places have none of these, so using the simple calgrow index seemed advisable.

Appendix 2
Summary of Past Studies on Growth Limitation

As we found relatively little past work, we conducted a review by checking items listed in the University of Minnesota Library's computer periodical index. From there, the search was extended backward by looking up articles they cited and forward by checking the articles they were cited in. In addition, recent issues of periodicals like the *Journal of Urban Affairs* were skimmed. A few recent items were added by Clark and Goetz. Jonathan Silverstein helpfully completed most of this search and a longer background memo.

Abbreviations

GC = growth control; Q = questionnaire; GE = growth effort; I = interviews; PH = public hearings; Pr = press reports; CC = central city; Sub = suburb; ICMA = International City Management Association Survey

Study	*Conclusions*	*Data*
Albrecht, Bultena, & Hoiberg	GC correlates negatively w/ income, education, occupation; businesspeople more progrowth than other white collar	Q from 465 Iowa residents
Anglin	Citizens who perceive their communities as rapidly growing are less likely to support more growth	I from 1000 NJ residents
Baldassare & Protash	GC not affect growth or satisfaction; income, race not predict GC; activism, white collar predict GC	Q 97 N. CA planning agencies, Q 321 N. CA residents
Dowall	GC communities are whiter, richer	74 ICMA Census of local gov'ts, census of agriculture
Frieden	GC supporters use environmental rhetoric, show no concern about spreading sprawl outside their community	Pr from 5 projects near SF
Garkovich	No controversy over plan's general goals; prop. owners, developers fight specific zoning	PH from Jessinme Co., KY, Q 650 residents
Gottdiener & Neiman	Liberals support GC, no economic status correlation	Q from 435 Riverside residents
B. M. Green & Schreuder	GC supporters much more likely to show up at zoning hearings after 1975; showing up had a much higher impact after 1975	PH from Wilmington, DE
G. P. Green & Fleischman	CCs have most GE, subs least; among subs GE relates positively to poverty	1984 ICMA; 1970, 1980 Census
Johnston	Taxes, prop. values, lifestyle motivate GC, economic exclusion doesn't	I, PH from Petaluma, Marin Co., Sacramento, Modesto
Logan & Zhou	GC has little effect on growth, pc income, rent, percent black	1973 ICMA; 1970, 1980 Census

Maurer & Christenson	Sub mayors less growth oriented; 1970-1975 pop growth not affect mayor attitude; higher per cap income = less growth-oriented mayor, but only in subs	I 325 KY mayors; 1970 Census, 1975 Census est.
Pogodzinski & Sass	Effect of commercial development on residential property values unclear	Literature review
Rips	No controversy over plan; controversy over "excessive" enforcement	I, Pr from Lincoln, NE
Vogel & Swanson	GC has broader appeal, GE has more resources: it is a close battle	I, PH, Pr from Gainesville, FL

Notes

1. For instance a Roper poll of the U.S. public ranked "environmentalism" Number 1 on a list of 20 salient items in July 1990 (*The American Enterprise* 1991, p. 88).

2. Elkin and Stone, as sensible political scientists, recognize variations in local leadership patterns and even introduce labels to capture some of these, such as Elkin's "pluralist," "entrepreneurial," and "federal political economies." Nevertheless, he writes: "land-use coalitions were the most important actors in the pluralist political economies" (p. 55), (citing Dahl and Banfield), albeit pluralists implied relative autonomy of land use decisions. Federalist cities develop(ed) more Washington-oriented policies. But Dallas and "all entrepreneurial cities have in common a relatively unimpeded alliance at work composed of public officials and local businessmen, an alliance that is able to shape the workings of city political institutions so as to foster economic growth" (p. 61). Stone: "the two sources I have found most compelling are Floyd Hunter's *Community Power Structure,* as an antecedent of preemptive power, and Stephen L. Elkin's formulation of urban regimes" (Stone 1989a: 145).

3. His assessment of experiences in many U.S. cities suggests that business dominance is the standard pattern, in that his six points in "Summing It Up" range from: "1. Business control of investment activity is a basic feature of all regimes . . ." to " 6. Unless public office holders are attached to a traditional pattern of small-stakes patronage or perhaps a new-fiscal-populism ideology, they are likely to be drawn toward an alliance with corporate interests" (1987, p. 287).

4. All items show the expected result, i.e., less responsive leaders generate more antigrowth movement, but not all are statistically significant in t-tests of differences between cities with and without antigrowth groups. Those significant at the .10 level are responsiveness to low-income groups, civic groups, local media, churches and religious groups, individual citizens, and taking a position unpopular with your constituency.

5. This can pit older residents against new, as does the "wolf issue" in Montana and nearby: older residents, especially ranchers, want wolves killed on the grounds that they attack cattle; the newer residents want to save the wolves ("Cry Wolf!" 1992).

6. Is this causality? It seems more plausible that antigrowth groups would increase antigrowth policies than vice versa. Still, both are surely in part explained by several more general underlying factors, as discussed here and in Chapter Appendix 1.

7. A slight correlation with one ICMA86 item in Table 4.4 disappeared in regressions.

References

Aiken, Michael and Paul E. Mott, eds. 1970. *The Structure of Community Power*. New York: Random House.

Albrecht, Don E., Gordon Bultena, and Eric Hoiberg. 1986. "Constituency of the Antigrowth Movement: A Comparison of the Growth Orientations of Urban Status Groups." *Urban Affairs Quarterly* 21 (4):607-16.

American Enterprise, The. 1991. "Popular Tastes: What's In, What's Out." (July/August):88-92.

Anderson, Martin. 1964. *The Federal Bulldozer*. Cambridge, MA: MIT Press.

Anglin, Ronald. 1990. "Diminishing Utility: The Effect on Citizen Preferences for Local Growth." *Urban Affairs Quarterly* 25 (4):684-96.

Appleton, Lynn M. and Terry Nichols Clark. 1989. "Coping in American Cities: Fiscal Austerity and Urban Innovation in the 1980s." Pp. 31-58 in *Urban Innovation and Autonomy*, edited by Susan E. Clarke. Beverly Hills, CA: Sage.

Atkinson, A. B. 1970. "On the Measurement of Inequality." *Journal of Economic Theory* 2:244-63.

Baldassare, Mark and William Protash. 1982. "Growth Controls, Population Growth and Community Satisfaction." *American Sociological Review* 47:339-46.

Bennett, Larry. 1986. "Beyond Urban Renewal: Chicago's North Loop Redevelopment Project." *Urban Affairs Quarterly* 22:242-60.

Bobo, Lawrence and Franklin D. Gilliam. 1990. "Race, Socio-Political Participation, and Black Empowerment." *American Political Science Review* 84:377-93.

Boyte, Harry. 1980. *The Backyard Revolution: Understanding the New Citizen Movement*. Philadelphia: Temple University Press.

Browning, Rufus P., Dale Rogers Marshall, and David H. Tabb. 1984. *Protest Is Not Enough*. Berkeley: University of California Press.

Clark, Terry Nichols. 1989. "What Causes Political Cultures?" Pp. 7-24 in *New Leaders, Parties and Groups*, edited by Harald Baldersheim, Richard Baldersheim, Richard Balme, Terry Nichols Clark, Vincent Hoffman-Martinot, and Hakan Magnusson. Paris and Bordeaux, France: CERVEL-IEP.

——— and Lorna Crowley Ferguson. 1983. *City Money: Political Processes, Fiscal Strain, and Retrenchment*. New York: Columbia University Press.

——— and Ronald Inglehart. 1990. "The New Political Culture." Prepared for the session on the New Political Culture, Research Committee 03, Community Research, World Congress, International Sociological Association, Madrid, July 9-14.

——— and Seymour Martin Lipset. 1991. "Are Social Classes Dying?" *International Sociology* 6, 4(December): 397-410.

Clarke, Susan E. and Gary L. Gaile. 1991. "The Next Wave: Post-Federal Local Economic Development Strategies." Presented to Chicago Conference on Leadership and Economic Development, July 22-24; forthcoming in *Economic Development Quarterly*.

——— and Michael J. Rich. 1985. "Making Money Work." Pp. 101-16 in *Coping With Urban Austerity, Research in Urban Policy*, edited by Terry Nichols Clark. Greenwich, CT: JAI.

Clavel, Pierre. 1986. *The Progressive City: Planning and Participation, 1969-1984*. New Brunswick, NJ: Rutgers University Press.

——— and Nancy Kleniewski. 1990. "Space for Progressive Local Policy: Examples From the United States and the United Kingdom." Pp. 192-234 in *Beyond the City Limits*, edited by John Logan and Todd Swanstrom. Philadelphia: Temple University Press.

"Cry Wolf!" 1992. *The Economist* 323, 7764 (June 20-26):91-92.

Dahl, Robert A. 1961. *Who Governs?* New Haven: Yale University Press.

Davis, Mike. 1990. *City of Quartz*. London: Verso.

DeLeon, Richard E. 1992. *Left Coast City: Progressive Politics in San Francisco, 1975-1991*. Lawrence: University Press of Kansas.

Delgado, Gary. 1986. *Organizing the Movement: The Roots and Growth of ACORN*. Philadelphia: Temple University Press.

DeSantis, Victor. 1991. Unpublished Ph.D. thesis and Conference Abstract. Chicago Conference on Leadership and Economic Development, July 22-24.

DiGaetano, Alan. 1990. "Urban Political Regime Formation: A Study in Contrast." *Journal of Urban Affairs* 11:261-82.

Donovan, Todd and Max Neiman. 1992. "Community Social Status, Suburban Growth, and Local Government Restrictions on Residential Development." *Urban Affairs Quarterly* 28(2):323-36.

———. Unpublished. "Citizen Mobilization and the Adoption of Local Growth Control." Department of Political Science, University of California, Riverside.

Dowall, David E. 1980. "An Examination of Population-Growth-Managing Communities." *Policy Studies Journal* 9:414-27.

Elazar, Daniel J. and Joseph Zikmund II, eds. 1975. *The Ecology of American Political Culture*. New York: Thomas Y. Crowell.

Elberling, John H. Unpublished report. "Community Origins of the Downtown Plan." Tenants and Owners Development Corporation, San Francisco, CA.

Elkin, Stephen L. 1987. *City and Regime in the American Republic*. Chicago: University of Chicago Press.

Fainstein, Norman I. and Susan S. Fainstein. 1974. *Urban Political Movements*. Englewood Cliffs, NJ: Prentice Hall.

Feagin, Joe R. and Michael Peter Smith. 1987. "Cities and the New International Division of Labor: An Overview." Pp. 3-34 in *The Capitalist City*, edited by Michael Peter Smith and Joe R. Feagin. Oxford, UK: Basil Blackwell.

Feiock, Richard. 1991. "The Effects of Economic Development Policy on Local Economic Growth." *American Journal of Political Science* 35(3):643-55.

Fisher, Robert. 1984. *Let the People Decide: Neighborhood Organizing in America*. Boston: Twayne.

Frieden, Bernard J. 1979. "The New Regulation Comes to Suburbia." *Public Interest* 55:15-27.

Gamson, William. 1990. *The Strategy of Social Protest.* Belmont, CA: Wadsworth.
Garkovich, Lorraine. 1982. "Land Use Planning as a Response to Rapid Population Growth and Community Change." *Rural Sociology* 47(1):47-67.
Goetz, Edward G. 1989. "Office-Housing Linkage in San Francisco." *Journal of the American Planning Association* 55:66-77.
———. 1990. "Type II Policy and Mandated Benefits in Economic Development." *Urban Affairs Quarterly* 26(2):170-90.
Gottdiener, Mark and Max Neiman. 1981. "Characteristics of Support for Local Growth Control." *Urban Affairs Quarterly* 17(1): 55-73.
Green, Brian M. and Yda Schreuder. 1991. "Growth, Zoning, and Neighborhood Organizations: Land Use Conflict in Wilmington, Delaware." *Journal of Urban Affairs* 13(1):97-110.
Green, Gary P. and Arnold Fleischman. 1991. "Promoting Economic Development: A Comparison of Central Cities, Suburbs, and Nonmetropolitan Communities." *Urban Affairs Quarterly* 27(1):145-54.
Hartman, Chester. 1984. *The Transformation of San Francisco.* Totowa, NJ: Rowman and Allanheld.
Hicks, Alexander, Roger Friedland, and Edward Johnson. 1978. *American Sociological Review* 43, 3(June):302-15.
Johnston, Robert A. 1980. "The Politics of Local Growth Control." *Policy Studies Journal* 9:428-40.
Kantor, Paul with Stephen David. 1988. *The Dependent City.* Glenview, IL: Scott, Foresman.
Logan, John R. and Min Zhou. 1989. "Do Suburban Growth Controls Control Growth?" *American Sociological Review* 54:461-71.
Logan, John R. and Harvey Molotch. 1987. *Urban Fortunes: The Political Economy of Place.* Berkeley: University of California Press.
Lowi, Theodore. 1979. *The End of Liberalism.* 2nd ed. New York: Norton.
Maurer, Richard C. and James A. Christenson. 1982. "Growth and Nongrowth Orientations of Urban, Suburban, and Rural Mayors: Reflections on the City as a Growth Machine." *Social Science Quarterly* 63(2):350-58.
Meador, Toni L. 1978. "A Proposed Development Management System for Boca Raton, Florida." FAU/FIU Joint Center for Environmental and Urban Problems.
Mollenkopf, John H. 1983. *Contested City.* Princeton, NJ: Princeton University Press.
Molotch, Harvey. 1976. "The City as a Growth Machine: Toward a Political Economy of Place." *American Journal of Sociology* 82:483-99.
O'Brien, David J. 1975. *Neighborhood Organization and Interest Group Processes.* Princeton, NJ: Princeton University Press.
Penne, R. Leo. 1976. "Boulder." *Nation's Cities* September 1986: 10-15.
Peterson, Paul E. 1981. *City Limits.* Chicago: University of Chicago Press.
Pogodzinski, J. M. and Tim R. Sass. 1991. "Measuring the Effects of Municipal Zoning Regulations: A Survey." *Urban Studies* 28(4):597-621.
Population-Environmental Balance, Inc. 1987. *Community Responses to Population Growth and Environmental Stress.* Washington, DC: Author.
Rips, Bruce P. 1991. "Managing Growth by the (Play) Book: Lincoln, Nebraska." *Urban Land* (April):21-25.
Rosdil, Donald. 1991. "The Context of Radical Populism in US Cities." *Journal of Urban Affairs* 13:77-96.

Saiz, Martin. 1990. "Determinants of Economic Development Policy Innovation Among the American States." Presented to International Sociological Association World Congress, Madrid, July 8-14.

Schiffman, Irving. 1990. *Alternative Technologies for Managing Growth.* Berkeley, CA: Institute of Governmental Studies.

Schmenner, Roger. 1982. *Making Business Location Decisions.* Englewood Cliffs, NJ: Prentice-Hall.

Schneider, Mark. 1991. "The Antigrowth Entrepreneur." Presented to the annual meeting of the American Political Science Association, Washington, DC.

———. 1992. "Undermining the Growth Machine: The Missing Link Between Local Economic Development and Fiscal Payoffs." *Journal of Politics* 54(1):214-30.

——— and Paul Teske. 1992. "Toward a Theory of the Political Entrepreneur: Evidence From Local Government." *American Political Science Review* 86(3):737-47.

Schumaker, Paul. 1991. *Critical Pluralism, Democratic Performance and Community Power.* Lawrence: University Press of Kansas.

Stone, Clarence. 1987. "Summing It Up." Pp. 269-90 in *The Politics of Urban Development,* edited by Clarence Stone and Heywood T. Saunders. Lawrence: University Press of Kansas.

———. 1989a. "Paradigms, Power and Urban Leadership." Pp. 135-59 in *Leadership and Politics,* edited by Bryan D. Jones. Lawrence: University Press of Kansas.

———. 1989b. *Regime Politics.* Lawrence: University Press of Kansas.

———, Marion Orr, and David Imbroscio. Forthcoming. "The Reshaping of Urban Leadership in US Cities." In a volume edited by Gottdiener and Pickvance.

Truman, David. 1951. *The Governmental Process.* New York: Knopf.

Turner, Robyne S. 1992. "Growth Politics and Downtown Development: The Economic Imperative in Sunbelt Cities." *Urban Affairs Quarterly* 28(1):3-21.

Vogel, Ronald K. and Bert E. Swanson. 1989. "The Growth Machine Versus the Antigrowth Coalition: The Battle for Our Communities." *Urban Affairs Quarterly* 25:63-85.

Webber, Carolyn and Aaron Wildavsky. 1986. *A History of Taxation and Expenditure in the Western World.* New York: Simon and Schuster.

Williams, John A. and Erwin Zimmermann. 1981. "American Business Organizations and Redistributive Preferences." *Urban Affairs Quarterly* 16(June):453-64.

Wong, Kenneth. 1989. "Toward a 'Political Choice' Model in Local Policy Making." Pp. 217-46 in *Decisions on Urban Dollars, Research in Urban Policy,* edited by Terry Nichols Clark, William Lyons, and Michael Fitzgerald. Greenwich, CT: JAI.

5

Growth and Decline of City Government

Rowan A. Miranda
Norman Walzer

Introduction

In this chapter, we examine the growth and decline of municipal expenditures during the 1970-1989 period. In Section I, six classes of hypotheses are discussed based on a review of the literature on growth of government. The empirical analysis is presented in two stages. In Section II, data from the Fiscal Austerity and Urban Innovation (FAUI) Project are used to test hypotheses from the literature on the growth of government. Because the FAUI data were collected in 1984, only changes in expenditures during 1980-1984 are examined. In the latter part of Section II, we use data from the NORC Permanent Community Sample (PCS) to examine the determinants of municipal expenditures at different points in the 1970-1989 period. Numerous cross-sectional analyses of expenditures have been conducted, but few have examined local government growth in a longitudinal design. Although analysis based on the FAUI data allows us to introduce and test more explanations for government growth, the longitudinal analysis on the smaller sample of PCS cities is neverthe-

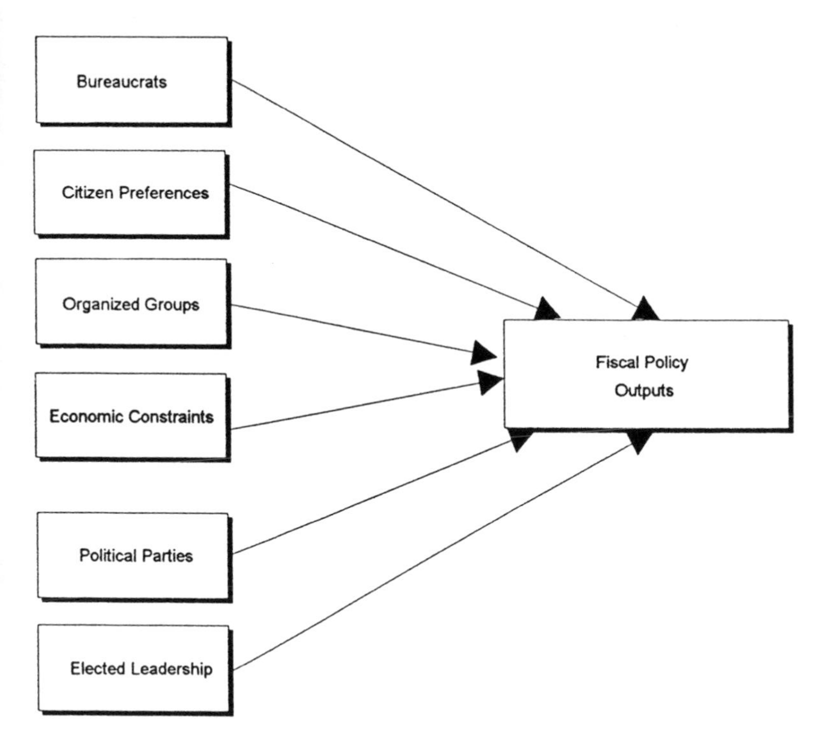

Figure 5.1. Explanations for Government Growth

less useful in assessing the relative importance of explanatory variables over time.

I. Theories of Public Sector Growth

This section builds on several reviews of the growth of government literature (Tarschys 1975; Cameron 1978; Larkey, Stolp, and Winer 1981; Wildavsky 1985). The number of theories about the growth of government exceed what can be manageably covered in any single research design. We attempt here to examine empirical support for the major explanations for government growth in this literature (Figure 5.1).

Propositions

1. Economic Constraints

Proposition 1: As local resources increase, governments will increase public expenditures.

This is a common explanation for government growth (Fabricant 1952; Bahl and Saunders 1965; Dye 1966; Hofferbert 1966); it is also the one that has received the most consistent empirical support (Boyne 1988). Resources include income, population, and intergovernmental aid. Like Wagner's Law of Increasing State Activity, Proposition 1 is essentially a statement of an income-elastic government. Under "good" times governments increase spending; retrenchment is a response to lower resources, such as a cut in intergovernmental aid or a decline in the local private sector resource base.

In *City Limits*, Peterson (1981) examines the determinants of local spending and observes: "Fiscal capacity remains the primary determinant of redistributive expenditures; demand-supply variables remain the most important determinants of developmental expenditures; and allocational policies remain affected by both fiscal capacity and demand-supply factors" (p. 58). For some time scholars debated the relative importance of political and economic variables (Sharpe and Newton 1984; Boyne 1988). Recent studies have moved past this futile debate by accepting that economic factors circumscribe the policy choices of local officials; these studies nevertheless contend that the choices of local officials remain significant within economic constraints (Clark and Ferguson 1983; Elkin 1987; Stone 1989; Wong 1988, 1990). Wong outlines a "political choice" framework that associates a strong flow of causation to economic constraints, but contends that "within each arena, noneconomic factors generate a wide range of policy alternatives which may or may not reinforce the thrust of the structural constraint for the city as a whole" (1988, p. 1). To some extent, the propositions below are factors that may displace or reinforce the predictions of economic constraint explanations.

The key measures of economic constraints examined in this study include levels and changes in population and income. (Unemployment was used in preliminary analyses.)

2. *Citizen Preferences*

Proposition 2: The larger the sector (of like citizens) in a city, the more fiscal policy is responsive to that sector.

Clark and Ferguson (1983) use the concept of "sector" to refer to citizens with like preferences concerning urban policies. The "middle class," "blacks" and "the poor" are potential analytical categories of citizens with like preferences. Downs's (1957) economic theory of democracy suggests that politicians pursue the policy preferences of the median voter by carrying out "those acts of spending which gain the most votes by means of those acts of financing which lose the fewest votes" (p. 52). The growth of a particular sector should lead to an increase in the responsiveness of government to the policy preferences of that sector.

The main measure of citizen preferences used in this study is percent black population.[1]

3. *Bureaucrats*

Proposition 3: The greater the power of bureaucrats vis-à-vis oversight authorities, the greater the expenditures.

In *Bureaucracy and Representative Government* (1971), Niskanen developed a theory of government bureau behavior based on the central assumption that bureaucrats attempt to maximize their budgets. Budget maximization allows bureaucrats to increase pay, power, and prestige. Under imperfect information, the relationship between a bureau and a sponsor (i.e., legislative committee) is one of "bilateral monopoly." Bureaucrats know the true cost of government output and exploit the informational asymmetries facing legislative oversight committees. The Niskanen theory thus predicts that rational bureaucrats will succeed in the expansion of the public sector beyond that desired by citizens. The FAUI survey asked mayors: "In the last three years, how important was the professional staff as compared to elected officials in affecting the overall spending level of your city government? How about allocating funds across departments? How about in developing new fiscal management strategies?" For each question, options listed vary from "elected officials largely decide" to "professional staff largely decide," which together with intermediate categories made up a five-point scale. An index of staff power was created by adding responses to the three questions; the higher the value on the staff power index, the greater the hypothesized impact on expenditure increases.

4. Organized Groups

Proposition 4: The greater the political power of (program-demanding) organized groups, the higher the expenditures.

The activities of organized groups can be traced to the work of pluralists (Truman 1951; Dahl 1961; Polsby 1980), to recent studies using regime analysis (Elkin 1987; Stone 1989) in urban politics and "rent seeking" in the public choice literature (Mueller 1989). In public choice theory, rents are economic resources extracted from government over and above the outcome that the market would generate. For an analysis of municipal expenditures, major "rent-seeking" organized groups are municipal employee unions. Politicians have an incentive to be responsive to unions because of their threat to strike and the votes they can garner.

Gramlich and Rubinfeld (1982) have shown that municipal employees have higher rates of turnout than other groups in local elections. Zax (1989) finds that cities with organized municipal employees have higher employment levels. L. Edwards and F. Edwards (1982) find that unionization increases compensation to municipal employees with effects ranging from 1 to 63 percent, "depending on the precise measure of employee compensation that is examined (money or fringe benefits)" and on the city's form of local government (p. 421). In particular, unions have an easier time extracting higher compensation from mayor-council cities relative to manager-council cities.

5. Political Parties

Proposition 5: Cities with mayors affiliated with the Democratic party will have higher expenditures.

Numerous studies have examined the linkage between political parties and public policy in different countries (Wilensky 1975; Castles

1978). In several studies of municipal expenditures in European countries, parties have been shown to impact fiscal policies (see Page, Goldsmith, and Kousgaard 1990). Studies of local government expenditures in the United States have generally not found differences between cities with Democrat and Republican mayors. Although many cities in the United States have nonpartisan elections, the FAUI survey nevertheless allowed local officials to choose a party affiliation. These responses are used to construct a dummy variable for Democrat mayors to analyze potential party effects on expenditure growth.

6. Elected Officials

Proposition 6: The more fiscally liberal the leadership, the higher the expenditures.

The political leaders examined in this study are mayors and council members. Clark and Ferguson (1983) suggest that political leaders can be "invisible" in the Downsian sense of strictly representing the preferences of the median voter or "dynamic," with policy preferences of their own (p. 25). Decisions about expenditure changes are mediated by council members and mayors, but the relative importance of each varies by city. An improved understanding of the importance of municipal officials in spending decisions is crucial in predicting expenditure changes, but detailed information on spending preferences of mayors and council members for large samples of cities has been lacking.

The FAUI questionnaire collected information on the expenditure preferences of mayors and council members on 13 areas (such as police, welfare, and so forth). Jones and Walzer (1988) studied the importance of council members' and mayors' attitudes toward spending in determining expenditure levels. This section draws heavily on that analysis. Possible responses to FAUI questionnaire items were to spend "a lot less," "somewhat less," "the same," "somewhat more," or "a lot more." Responses to the 13 categories were used to create indices for the fiscal liberalism of the mayor and council members by adding responses across the 13 categories. Table 5.1 shows results of a cross tabulation between spending preferences of mayors against those of council members. Although the modal category was consensus among both to decrease spending (41 cities in

TABLE 5.1 Expenditure trends in U.S. Cities, 1980-1984

		Per Capita Expenditures		
Expenditure Categories		*1980($)*	*1984($)*	*Percent Change*
General Expenditures	Nominal	432.06	552.75	27.9
	Real	432.06	441.49	2.2
Current Operations	Nominal	319.91	416.53	30.2
	Real	319.91	332.68	4.0
Capital	Nominal	84.86	85.53	0.8
	Real	84.86	68.31	–19.5
Common functions	Nominal	238.31	312.04	30.9
	Real	238.31	249.23	4.6
Highways	Nominal	42.94	53.62	24.9
	Real	42.94	43.52	1.4
Fire	Nominal	37.23	50.99	37.0
	Real	37.23	42.14	13.2
Financial Administration	Nominal	9.67	14.23	47.2
	Real	9.67	11.21	15.9
Police	Nominal	50.74	72.82	43.5
	Real	50.74	68.10	34.2
Libraries	Nominal	9.28	12.07	30.1
	Real	9.28	9.22	–0.6
Parks and Recreation	Nominal	25.59	33.45	30.7
	Real	25.59	25.69	0.4
Sanitation	Nominal	18.04	23.54	30.5
	Real	18.04	18.85	4.5
Sewerage	Nominal	41.42	50.65	22.3
	Real	41.42	38.37	–7.4

SOURCE: U.S. Bureau of the Census (various years).
NOTE: Common functions expenditures include the eight functions listed below it as well as general public buildings, which is not shown.

Table 5.1), in more than half the cities there was a moderate to high degree of disagreement. Our analysis allows us to examine whether the preferences of the mayor, the council, or both are influential in explaining spending patterns.

II. Empirical Analysis

In this section, we present results from multiple regression equations of municipal expenditures. First, however, it is useful to describe the fiscal climate during this period.

Fiscal Climate

City expenditures are determined by service responsibilities mandated in state constitutions and statutes. The position of cities under U.S. federalism leads them to provide a similar set of services with special functions financed by other public bodies, such as special districts. Yet there are differences in service responsibilities that affect expenditure changes in several ways. Cities responsible for more services reportedly have larger compensation increases (Clark and Ferguson 1983, p. 162). Personnel costs are a major proportion of city expenditures, and therefore salary increases translate into higher expenditure growth. Cities responsible for services such as health and welfare, for which the costs have increased rapidly, may face more fiscal pressure and face trade-offs among service categories. Federal and state aid finance local services; differences in service responsibility allow differential access to this aid. For these and other reasons, comparisons of expenditure changes must be adjusted for variations in responsibility for services.

For this study, we adopt the common functions approach in comparing expenditures. Common functions expenditures include the following services: police, fire, highways, sewerage, sanitation, general buildings, parks and recreation, general control, and financial administration (Clark and Ferguson 1983, p. 46). These services represent approximately 64.7 percent of total general expenditures by cities. Services excluded are welfare, hospitals, libraries, and education, which are sometimes provided by single-purpose districts, counties, or even state governments.

Nationally, U.S. cities larger than 25,000 spent $312.04 per capita on common functions in 1984, an increase of 30.9 percent from the $238.31 per capita spent in 1980 (Table 5.2). Inflation—double digit during several years—markedly raised the costs of providing city services and make current dollar comparisons misleading. In constant dollars, city expenditures increased only 5 percent.[2]

Current operations increased 4 percent in 1980 dollars between 1980 and 1984, but per capita capital expenditures declined 19.5 percent. City officials apparently postponed capital expenditures to preserve current spending. Even then, per capita spending increased only 5 percent between 1980 and 1984.

Resources were also reallocated among service categories between 1980 and 1984. Per capita expenditures increased most in financial

TABLE 5.2 Conflict and Consensus on Spending Preferences

		Mayor Preferences (Number of Cities)			
		Increase	*Same*	*Decrease*	Totals
Council Preferences	*Increase*	18	16	6 (high conflict)	40
	Same	27	10	17	54
	Decrease	6 (high conflict)	15	41	62
	Totals	51	41	64	156

SOURCE: Fiscal Austerity and Urban Innovation Project.

administration, police, and fire protection, with the smallest nominal increases in sewerage treatment and highways. The smallest real increases were in parks and recreation, followed by highways and sanitation. Smaller increases in parks/recreation and sanitation than in police and fire protection are not unexpected in a tight fiscal environment. Delayed or reduced recreation expenditures are less threatening to residents than cutting police and fire protection, which are considered emergency services.

Model Specification

Three separate regressions are estimated to identify determinants of expenditure changes with the following regression model estimated for each dependent variable:

$$Y = B_0 X_1^{B1} X_2^{B2} U \text{ where } B_1 = e^{B1} \text{ and } U = e_u$$

which is estimated as

$$\log Y = \log B_0 + \sum_{i=1}^{11} b_i \log X_i + \log u_i ,$$

where X_i represents:

X_1 = City population in 1980
X_2 = Ratio of city population in 1984 to 1980
X_3 = Ratio of city per capita income in 1983 to 1979
X_4 = Ratio of city unemployment rate in 1982 to 1980
X_5 = Ratio of tax effort in 1984 to 1980
X_6 = Ratio of total intergovernmental revenue in 1984 to 1980
X_7 = Staff power index
X_8 = Party of the mayor
X_9 = Expenditure preferences of the mayor
X_{10} = Expenditure preferences of the majority council member
X_{11} = Percent organized municipal employees

We consider this model first using FAUI data in the 1980s and next using NORC-PCS data for 1970-1989. (The National Opinion Research Center-Permanent Community Sample is a set of 62 cities resurveyed multiple times; it is the main data source of Clark and Ferguson 1983.)

Independent Variables. Previous expenditure determinant studies include population, population change, income, and poverty as independent variables, so these variables will not be explained in detail here (Clark and Ferguson 1983, p. 136). Wealthy cities are expected to adopt larger spending increases or at least maintain expenditures during a recessionary period. Those with high unemployment and more poverty may report smaller expenditure increases or even declines. Cities with greater population growth may have greater expenditure increases than those with economic decline. The ratio of intergovernmental revenue in 1984 to 1980 is also examined. Intergovernmental revenue consists of aid from all sources including federal, state, and other local governments. Federal grants have General Revenue Sharing removed.

Dependent Variables. Determinants of common function expenditure levels in 1984 are reported first followed by the common function expenditures between 1980 and 1984. Following Clark and Ferguson (1983), we conceptualize fiscal strain as an imbalance between fiscal policies and private sector resources. Consequently, we introduce a third dependent variable to assess this—change in common functions divided by changes in per capita income. Ratios between 1980 and 1984 are used rather than percentage changes so that the variables can be expressed in logs.

Regression Results. Table 5.3 presents regression results of common function expenditure levels and changes. Intergovernmental aid, income, and population are positively associated in explaining common function expenditure levels. Also, the greater the fiscal liberalism of the majority council members, the higher the common function expenditure levels. Insignificant are staff power, citizen preferences, organized municipal employees, party of the mayor, or the mayor's fiscal liberalism.

Examining changes in common function expenditures from 1980 to 1984, we find that these are higher in cities with more population growth. Similarly, the more fiscally liberal the majority council members, the higher the expenditures. Once again, the other hypotheses did not receive empirical support.

The regression results on the fiscal strain measure—changes in common function expenditures to change in median income between 1980 and 1984—demonstrate that only council fiscal liberalism is statistically significant: the more fiscally liberal the city council, the greater the fiscal strain.

In summary, the Table 5.3 results support the economic constraint and elected leadership hypotheses most consistently. The R^2 for all three equations are modest, with the variance explained for expenditure levels being considerably higher than that for expenditure changes.

We next replicate this analysis in two sets of cities: Tables 5.4 and 5.5 separate mayor-council cities from council-manager cities. Comparison of the two tables shows an interaction effect between form of government and importance of council preferences. In council-manager cities, council preferences are significant for explaining changes in common functions, and common functions change to per capita income change (Table 5.4). Both the slopes and standardized regression coefficients for council fiscal liberalism in council-manager cities are higher than that of the overall sample (Table 5.3). By

TABLE 5.3 Determinants of Common Functions Expenditures, 1980-1984

	Level of Common Functions 1984			*Ratio Common Functions 1984/1980*			*Ratio of CF 84/80 to Income 84/80*		
Variables	*B*	*Beta*	*T*	*B*	*Beta*	*T*	*B*	*Beta*	*T*
Change in Population 1984/1980	—	—	—	1.380	.255	2.91**	.531	.104	1.18
Change in Income 1985/1980	—	—	—	.105	.015	.18	—	—	—
Change in Intergovernmental Aid	—	—	—	.186	.136	1.70	.184	.142	1.72
Intergovernmental Aid 1984	.119	.286	3.28**	—	—	—	—	—	—
Income 1980	.238	.162	1.96*	—	—	—	—	—	—
Population 1980	.108	.296	3.52**	.011	.047	.53	.011	.050	.54
Percent Black 1980	.186	.059	.70	–.266	–.126	–1.39	–.262	–.132	–1.40
Staff Power Index	.084	.135	1.89	.025	.061	.76	.024	.064	.77
Organized Municipal Employees	.135	.080	1.04	–.081	–.074	–.89	–.085	–.082	–.95
Democrat Mayor Dummy	–.034	–.051	–.69	–.015	–.034	–.41	–.019	–.047	–.57
Mayor Fiscal Liberalism Index	.040	.024	.33	–.116	–.109	–1.31	–.128	–.127	–1.49
Council Fiscal Liberalism Index	.339	.168	2.24**	.321	.247	2.96**	.313	.257	2.97**
Constant	1.187		1.02	–.058		–.20	.085		.32
R Square		.32			.20			.13	
Adjusted *R* Square		.27			.14			.07	
Number of Cases		151.00			151.00			151.00	

SOURCE: Fiscal Austerity and Urban Innovation Project.
NOTE: Regression equations are log-log models, except dummy variables. * = $p < .05$; ** = $p < .025$.

TABLE 5.4 Common Functions Expenditures in Council-Manager Cities, 1980-1984

	Level of Common Functions 1984			*Ratio Common Functions 1984/1980*			*Ratio of CF 84/80 to Income 84/80*		
Variables	*B*	*Beta*	*T*	*B*	*Beta*	*T*	*B*	*Beta*	*T*
Change in Population 1984/1980	—	—	—	1.678	.299	2.42**	.966	.185	1.52
Change in Income 1985/1980	—	—	—	.966	.098	.82	—	—	—
Change in Intergovernmental Aid	—	—	—	.167	.100	.90	.153	.099	.87
Intergovernmental Aid 1984	.072	.167	1.34	—	—	—	—	—	—
Income 1980	.139	.101	.81	—	—	—	—	—	—
Population 1980	.114	.260	2.07*	.026	.077	.62	.023	.074	.58
Percent Black 1980	.210	.056	.41	–.607	–.209	–1.67	–.574	–.212	–1.66
Staff Power Index	.055	.090	.85	.025	.054	.51	.026	.058	.52
Organized Municipal Employees	.151	.086	.71	–.018	–.013	–.12	–.013	–.010	–.09
Democrat Mayor Dummy	–.051	–.076	–.70	–.039	–.075	–.67	–.031	–.064	–.58
Mayor Fiscal Liberalism Index	.215	.106	1.01	–.294	–.189	–1.76	–.280	–.193	–1.76
Council Fiscal Liberalism Index	.434	.224	2.03*	.397	.267	2.47**	.383	.276	2.46**
Constant	1.940		1.16	–.294		–.51	.025		.05
R Square		.26			.29			.21	
Adjusted *R* Square		.16			.18			.10	
Number of Cases		81.00			81.00			81.00	

SOURCE: Fiscal Austerity and Urban Innovation Project.
NOTE: Regression equations are log-log models, except dummy variables. * = $p < .05$; ** = $p < .025$.

TABLE 5.5 Common Function Expenditures in Mayor-Council Cities, 1980-1984

	Level of Common Functions 1984			*Ratio Common Functions 1984/1980*			*Ratio of CF 84/80 to Income 84/80*		
Variables	*B*	*Beta*	*T*	*B*	*Beta*	*T*	*B*	*Beta*	*T*
Change in Population 1984/1980	—	—	—	.171	.032	.23	–.550	–.104	–.75
Change in Income 1985/1980	—	—	—	–.204	–.047	–.34	—	—	—
Change in Intergovernmental Aid	—	—	—	.132	.144	1.04	.088	.097	.72
Intergovernmental Aid 1984	.219	.509	4.17**	—	—	—	—	—	—
Income 1980	.362	.206	1.95*	—	—	—	—	—	—
Population 1980	.073	.230	1.92	.022	.162	1.06	.017	.125	.83
Percent Black 1980	.327	.119	1.07	–.089	–.070	–.46	–.065	–.051	–.34
Staff Power Index	.093	.142	1.46	.018	.061	.44	.026	.091	.67
Organized Municipal Employees	.082	.049	.48	–.173	–.236	–1.67	–.189	–.259	–1.85
Democrat Mayor Dummy	–.021	–.032	–.30	–.029	–.098	–.66	–.031	–.107	–.72
Mayor Fiscal Liberalism Index	–.054	–.038	–.36	.013	.021	.14	–.010	–.017	–.12
Council Fiscal Liberalism Index	.200	.093	.84	.079	.084	.53	.065	.070	.44
Constant	.243		.14	.228		.78	.396		1.39
R Square		.47			.12			.11	
Adjusted *R* Square		.40			–.05			–.04	
Number of Cases		70.00			70.00			70.00	

SOURCE: Fiscal Austerity and Urban Innovation Project.
NOTE: Regression equations are log-log models, except dummy variables. * = $p < .05$; ** = $p < .025$.

contrast, in mayor-council cities, mayor and council fiscal liberalism are insignificant. The only significant result in the three equations is that more intergovernmental aid and higher income increase common function expenditure levels. Thus, we do not find support for the economic constraint or any of the other hypotheses previously discussed in mayor-council cities (Table 5.5).

To summarize, during the 1980-1984 recessionary period, city governments were unable to respond to the policy preferences of various actors in the political environment. To the extent that the findings support any of the above propositions, it is the economic constraint and elected leadership hypotheses. The most consistent finding was that fiscal liberalism of the majority council members led to higher expenditure levels and changes, especially in council-manager cities.

Regressions were also computed for the 1985-1989 period on general expenditures and common functions. Although the models were not fully comparable to the 1980-1984 period because of data limitations, results were not markedly different from those for 1980-1984. Although council fiscal liberalism was not as discernible as in 1980-1984, economic constraints such as intergovernmental aid and income remained significant. The 1985-1989 period is examined using a smaller sample of cities for which we had data in Tables 5.6 and 5.7.

Are city governments more responsive to political pressures during periods with less fiscal strain? Results from our longitudinal study of expenditures over two decades are presented next.

Longitudinal Analysis of Expenditure Levels

We used data from PCS studies (explained under "Model Specification") to examine determinants of common function expenditure levels. Table 5.6 presents the results. The "FP Index" adjusts for differences in responsibilities across cities. Although this index is not as important to the analysis of "common" function expenditures, it does allow us to control for a larger service burden for cities with higher FP scores, which can mean they have less to spend on core city services. The most striking finding is the difference in the importance of percent black between the 1970s and 1980s. In both the 1970-1974 and 1975-1979 periods, a larger sector of blacks led to an expansion of city expenditures. By contrast in the 1980s, this sector did not have a statistically significant effect.

TABLE 5.6 Common Functions Expenditure Levels, 1970-1989

Variables	*Per Capita CF* 1970			*Per Capita CF* 1975			*Per Capita CF* 1980			*Per Capita CF* 1985			*Per Capita CF* 1989		
	B	*Beta*	*T*	*B*	*Beta*	*T*	*B*	*Beta*	*T*	*B*	*Beta*	*T*	*B*	*Beta*	*T*
Population	–.002	–.005	–.05	.025	.090	.91	.004	.016	.12	.035	.135	.94	.036	.120	.89
Per Capita Income	.753	.353	3.65**	.957	.486	5.77**	.709	.465	3.35**	.690	.487	3.29**	.959	.608	4.41**
Intergovernmental Aid	.162	.515	4.49**	.287	.739	5.82**	.125	.338	2.16*	.065	.182	1.06	.078	.228	1.46
FP Index	.035	.115	.94	–.028	–.101	–.80	.153	.237	1.65	.056	.085	.55	–.023	–.030	–.21
City Age	.107	.180	1.37	.007	.013	.11	.060	.116	.68	.119	.228	1.24	.141	.235	1.35
Council-Manager Form	.005	.007	.09	–.028	–.043	–.51	–.139	–.242	–1.90*	–.119	–.206	–1.49	–.018	–.027	–.21
Percent Black	.060	.209	1.99**	.055	.206	2.05*	.034	.165	1.26	.015	.071	.49	.047	.197	1.43
Constant	–2.900		–1.69	–4.395		–3.12**	–2.647		–1.29	–2.243		–1.09	–4.466		–2.09*
R Square		.67			.70			.47			.38			.44	
Adjusted *R* Square		.63			.66			.39			.29			.36	
Number of Cases		63.00			63.00			57.00			57.00			57.00	

SOURCE: NORC Permanent Community Sample.
NOTE: Regression equations are log-log models, except dummy variables. * = $p < .05$; ** = $p < .025$.

The variance explained by the model in the 1970s is almost twice that in the 1980s (Table 5.6). These results are for levels. What of changes?

Table 5.7 shows changes in expenditures for 5-year periods from 1970 to 1989. Once again, the most striking result is the role of the black sector. Changes in percent black led to increases in 5-year measures of common function change during the 1970s but *not* the 1980s. Indeed, in 1985-1989, cities with more black residents increased spending less, whereas larger, wealthier, and council-manager cities increased more than other cities. These results support the economic constraint hypothesis for the 1985-1989, but not in earlier periods.[3]

The explained variance of the change model is modest until 1985-1989. The higher slopes of percent black and changes in income are likely responsible for the increase in explained variance during this period.

Conclusions

In this chapter, we examined the growth of municipal expenditures over a period of nearly two decades. Six classes of hypotheses were derived from the literature on the growth of government. These propositions were tested in regression analyses of expenditure levels and changes during the 1970-1989 period. In particular, data from the FAUI survey were analyzed for 1980-1984. Data from the NORC Permanent Community Sample were used for the longitudinal analysis of the entire 1970-1989 period.

Some of our results support the economic constraint hypothesis, which suggests that cities with greater fiscal capacity spend more. Yet the role of economic constraints varies by period. Our results also point to the importance of council member fiscal liberalism in increasing expenditures, especially in council-manager cities. By contrast we found little support for hypotheses concerning bureaucrats, organized groups, and parties.

A longitudinal study of common functions expenditure levels and changes was conducted over a two-decade period to assess whether the fiscal behavior of governments differed in nonrecessionary times. The most striking finding was the importance of the black sector in increasing expenditures during the 1970s, although not in the 1980s.

TABLE 5.7 Common Functions Expenditure Changes, 1970-1989

Variables	*Change in per Capita CF 1974/1970*			*Change in per Capita CF 1979/1975*			*Change in per Capita CF 1984/1980*			*Change in per Capita CF 1989/1985*		
	B	*Beta*	*T*	*B*	*Beta*	*T*	*B*	*Beta*	*T*	*B*	*Beta*	*T*
Population	.051	.334	1.57	.004	.022	.11	.022	.210	1.24	.025	.275	2.22*
Change in Income	–.059	–.029	–.16	.133	.060	.36	–.019	–.011	–.07	.669	.592	5.02**
Change in Intergovernmental Aid	.044	.166	1.08	–.120	–.202	–1.48	.018	.030	.22	.038	.053	.52
FP Index	–.028	–.182	–.94	–.024	–.143	–.80	–.098	–.369	–2.11*	–.053	–.229	–1.76
City Age	–.029	–.073	–.32	.223	.516	2.44**	.021	.101	.52	–.046	–.250	–1.77
Council-Manager Form	–.050	–.145	–.87	.132	.350	2.27*	.067	.285	1.70	.063	.309	2.68**
Change in Percent Black	.048	.325	2.00*	.058	.357	2.42**	–.090	–.010	–.06	–2.683	–.340	–2.87**
Constant	–.175		–.39	–.781		–1.62	.833		.55	–2.130		–2.33**
R Square		.17			.29			.18			.56	
Adjusted *R* Square		.04			.18			.07			.50	
Number of Cases		53.00			53.00			57.00			57.00	

SOURCE: NORC Permanent Community Sample.
NOTE: Regression equations are log-log models, except dummy variables. *=p<.05; ** = p<.025.

The impact of minorities on government growth deserves comment. First, our results have limitations because we did not examine redistributive expenditures. However, the fact that we did not do so does not really explain the distinctive impact of the black sector between the 1970s and 1980s.

One explanation for the differences in the results for the two periods is the potentially greater activity of this sector during the early 1970s. In contrast, by the early 1980s, Clark and Ferguson (1983) argue that as "organized groups became less critical, leaders appealed to broader electoral constituencies, including whites as well as middle class blacks, who were less visibly united with poor blacks than a decade earlier" (p. 143). Another explanation is that "protest" by organized groups in the late 1960s and early 1970s was slowly replaced by minority incorporation into city councils. The 1980-1984 findings on the importance of council members in influencing expenditures may suggest that the direct effect of minority incorporation and not the relative size of the black sector then became a more important determinant of policy responsiveness. This result, if corroborated, would support the "protest is not enough" thesis that policy responsiveness to minorities occurs only after they are incorporated into city councils (Browning, Marshall, and Tabb 1984).

Finally, one of the most consistent findings was the role of council members in explaining fiscal policy outputs. Frequently, studies of U.S. cities proceed with a mayor-dominated model of policy making. More recently, scholars have paid more attention to city councils in urban politics (Welch and Bledsoe 1988). Findings from this study suggest that greater attention should be paid to the role of council members in influencing city budgetary decision making.

Why do we not find more support for the several theories outlined at the beginning of the chapter? Three reasons seem most likely: (a) the data and measures we have available may be too insensitive to capture more subtle workings of some theorized processes; (b) some theories may hold for a few cities, but too few to generate results of statistical significance in these national samples; and (c) the theories may be overstated and too simplified. They often stress a single-factor explanation, whereas life in all governments is multicausal and complex. Sometimes one factor drives policy, other times another factor does, in such a manner that we do not find a single clear result across time periods, but different variables operating in different periods.

Notes

1. Other measures of sectors were examined, such as percent middle class, percent elderly, and percent educated, but were found to be statistically insignificant in explaining expenditures during this period.

2. A fixed-weight price index was used to adjust expenditure changes for inflation. This index is a weighted aggregate of price relatives Lespayres index with 1977 (base year) weights. Price changes for goods and services particularly appropriate to reflect city purchases are included in the index. A separate index is computed for each major service provided by cities. Expenditure patterns based on a sample of Illinois cities larger than 25,000 are the base of the price index. See Walzer (1985).

3. Equations were examined that included percent black instead of *changes* in percent black for Table 5.7, to examine the hypothesis that the greater the absolute size of the sector the higher the expenditures. No support was found for this proposition.

The high coefficient for change in percent black for 1989/1985 led us to sensitivity testing of these results: several similar but slightly different models were specified to see if the results remained robust. The variables with high regression coefficients in 1989/1985 were reviewed for their simple *r*s with one another and with the dependent variable. Most showed the same strong *r*s with the dependent variable, although percent black was only $r = .27$. Next those with moderate interrelations—percent black, population, council-manager government, and city age—were removed one at a time from the model, to see if other coefficients changed. They did not. In particular, percent black remained strong with each of these other variables removed.

References

Bahl, R. and R. Saunders. 1965. "Determinants of Change in State and Local Government Expenditures." *National Tax Journal* 14:349-55.

Boyne, G. 1988. "Review Article: Theory, Methodology and Results in Political Science—The Case of Output Studies." *British Journal of Political Science* 15:473-515.

Browning, R., D. Marshall, and D. Tabb. 1984. *Protest Is Not Enough*. Berkeley: University of California Press.

Cameron, D. 1978. "The Expansion of the Public Economy: A Comparative Analysis." *American Political Science Review* 72, 4(December):1250-60.

Castles, F. 1978. *The Social Democratic Image of Society*. London: Routledge and Kegan Paul.

Clark, T. N. and L. Ferguson. 1983. *City Money: Political Processes, Fiscal Strain and Retrenchment*. New York: Columbia University Press.

Dahl, R. 1961. *Who Governs?* New Haven, CT: Yale University Press.

Downs, A. 1957. *An Economic Theory of Democracy*. New York: Harper & Row.

Dye, T. 1966. *Politics, Economics, and the Public*. Chicago: Rand McNally.

Edwards, L. and F. Edwards. 1982. "Public Unions, Local Government Structure and the Compensation of Municipal Sanitation Workers." *Economic Inquiry* 20(3):405-25.

Elkin, S. 1987. *City and Regime in the American Republic*. Chicago: University of Chicago Press.

Fabricant, S. 1952. *The Trend of Government Activity in the United States Since 1900.* New York: National Bureau of Economic Research.

Gramlich, E. and D. Rubinfeld. 1982. "Voting on Public Spending." *Journal of Policy Analysis and Management* 1(4):516-33.

Hofferbert, R. 1966. "The Relation Between Public Policy and Some Structural and Environmental Variables in the American States." *American Political Science Review* 50:73-82.

Jones, W., and N. Walzer. 1988. "Spending Patterns in American Cities." Unpublished manuscript.

Larkey, P., C. Stolp, and M. Winer. 1981. "Strategy Choices and Policy Outcomes in U.S. Cities," presentation to American Political Science Association meetings, Washington, D.C. (Mimeographed.)

Larkey, P., C. Stolp, and M. Winer. 1981. "Theorizing About the Growth of Government: A Research Assessment." *Journal of Public Policy* 1,2 (May):157-220.

Mueller, D. 1989. *Public Choice II.* Cambridge: Cambridge University Press.

Niskanen, W. 1971. *Bureaucracy and Representative Government.* Chicago: Adeline.

Page, E., M. Goldsmith, and P. Kousgaard. 1990. "Time Parties and Budgetary Change: Fiscal Decisions in English Cities, 1974-88." *British Journal of Political Science* 20:43-61.

Peterson, P. 1981. *City Limits.* Chicago: University of Chicago Press.

Polsby, N. 1980. *Community Power and Political Theory.* New Haven, CT: Yale University Press.

Sharpe, L. J. and K. Newton. 1984. *Does Politics Matter?* Oxford: Clarendon Press.

Stone, C. 1989. *Regime Politics.* Lawrence: University of Kansas Press.

Tarschys, D. 1975. "The Growth of Public Expenditures: Nine Modes of Explanation." *Scandinavian Political Studies* 10:9-31.

Truman, D. 1951. *The Governmental Process.* New York: Knopf.

U.S. Bureau of the Census. Various years. *City Government Finances.* Washington, DC: U.S. Government Printing Office.

Walzer, N. 1985. "1984 Municipal Price Index." *Illinois Municipal Review* (April: 13-15).

Welch, S. and T. Bledsoe. 1988. *Urban Reform and Its Consequences.* Chicago: University of Chicago Press.

Wildavsky, A. 1985. "The Logic of Public Sector Growth." Pp. 231-67 in *State and Market,* edited by Jan-Erik Lane. Beverly Hills, CA: Sage.

Wilensky, H. 1975. *The Welfare State and Equality.* Berkeley: University of California Press.

Wong, K. 1988. "Economic Constraint and Political Choice in Urban Policymaking." *American Journal of Political Science* 32,1 (February):1-18.

———. 1990. *City Choices: Education and Housing.* Albany: State University of New York Press.

Zax, J. 1989. "Employment and Local Public Sector Unions." *Industrial Relations* 28(1):21-31.

PART THREE

Coping in Lean Years: Making Fundamental Policy Changes

6

Cuts, Cultures, and City Limits in Reagan's New Federalism

Cal Clark
Oliver Walter

THE LATE 1970S AND EARLY 1980S presented a fiscal crisis unmatched since the Great Depression for all levels of government in the United States. Inflation, recession, and the concomitant structural and regional transformations of the U.S. economy took a considerable financial toll as governments faced declining revenues and increasing demands for public services. Thus, the era of ever-expanding budgets that had marked most of the postwar era passed into the

AUTHORS' NOTE: Edgar Bueno, Daniel Crane, Ed Harper, and Ziad Munson were students at the University of Chicago and research assistants with the Fiscal Austerity and Urban Innovation Project who helped update the tables as new fiscal data became available. A previous draft of this chapter was presented at the Annual Meeting of the American Political Science Association, Washington, D.C., September 1-4, 1988. We very gratefully acknowledge the detailed and valuable comments of Terry Nichols Clark, Robert M. Stein, and Kenneth K. Wong on earlier drafts. They should not, however, be held accountable for the inevitable shortcomings in our analysis.

dustbin of history; "cutback management" became the byword for many governments.

U.S. cities faced perhaps the severest fiscal crisis among U.S. governments. Cities must cope with many welfare and service demands generated by economic distress even as economic downturns and suburban flight have had strongly adverse consequences for their coffers. Thus, it is not surprising that fiscal austerity measures first appeared at the urban level, well before the federal budget crisis and even the much publicized "taxpayers' revolt" (T. N. Clark and Ferguson 1983).

Moreover, cities (and other local governments) are rather ill suited in their position in the federal fiscal system. Their tax bases are much narrower and less flexible than those of the state and federal governments. This had been offset by a system of intergovernmental transfers that accounted for a considerable proportion of city money. However, in response to the fiscal crisis of the early 1980s, federal and state governments curtailed their support of lower level politics to meet their own fiscal problems. Thus, the major federal tax cuts of the Reagan revolution were soon followed by tax increases by most states and many local governments. Unfortunately for the cities, there are no lower levels of governments to which their financial problems can be passed down. Thus, the Reagan administration's New Federalism of pushing responsibility for public service delivery downward to be "closer to the people" came at precisely the time that cities' ability to perform old, not to mention new, functions, appeared questionable.

Thus, one might well have expected an urban financial implosion in the mid-1980s as the gap between presumed resources and added responsibilities widened. Yet, a new crisis in city finances did not come until the post-Reagan recession. Scholarly evaluations of changes in the nature of U.S. intergovernmental transfer system during the Reagan administration, many based on detailed analysis of individual programs, have generally reached the paradoxical conclusions that considerable cuts were made, especially in federal funds to local governments, but that these cuts did not have a profound effect on local budgets for a variety of reasons. First, the cuts were not as severe as first feared; and after 1982 overall federal aid stabilized and even grew slightly. Second, the economic recovery that began in 1983 helped. Third, during the rapid expansion of federal aid in the 1960s and 1970s, few new funds had been substituted for financing basic state and local services, so that the need to find replacement funding was attenuated. Fourth, the states did provide considerable replacement funds for some programs, and many local governments responded to the cuts by significantly increasing their own local revenues, especially through tapping new sources such as users' fees (Appleton and T. N. Clark 1989; Cole 1990; Marando 1990; Nathan and Doolittle 1985; Nathan, Doolittle, and Associates 1987; Nathan and Lago 1988; G. E. Peterson 1984; P. E. Peterson, Rabe, and Wong 1986; Wolman 1986; Wong and P. E. Peterson 1986). Fifth, despite the general view that the Reagan revolution had produced a much more conservative federal policy, at least one sophisticated study concluded that considerable targeting of "needy" cities continued (Morgan and Shih 1991). Finally, economic pressures forced cities to become much more entrepreneurial in promoting economic development and devising innovative "privatization" projects in collaboration with business (Bueno 1992; Clarke and Gaile 1991; Heilman and Johnson 1992). Until the recession of the 1990s, these factors proved sufficient to keep most cities afloat. In contrast, with the post-Reagan slowdown, the "economic merry-go-round" stopped, and the specter of fiscal and social collapse in U.S. urban centers emerged once more.

From another perspective, however, federal cuts in intergovernmental aid would be expected to have had important social consequences. Paul Peterson (1981) has differentiated urban policies into "redistributional" support for the poor, "developmental" efforts to promote growth, and "allocative" policies involving basic services. Peterson argues that because of the competitive nature of city developmental efforts only the federal government has the interest or capacity to support redistributional policies. Thus, shifting former federal fiscal responsibilities to the states and cities should disproportionately cut social services for the poor (see Wong 1988 for an extension of this model to include political variables).

Analysis of the Reagan cuts suggests that this was true. First, the Reagan administration cut many entitlements and social programs despite the recession of the early 1980s (Danziger and Feaster 1985; Nathan and Lago 1988; Seymour et al. 1982). Second, for intergovernmental aid itself, state and city replacement efforts were almost exclusively limited to development and infrastructure programs rather than social services—health formed an exception explicable by the political power of the "service providers" in this area (Nathan and Doolittle 1985; Nathan, Dearborn, Goldman, and Associates 1982; G. E. Peterson 1984). Third, the form of federal aid changed significantly. The administration's consolidation of categorical grants into much broader block grants decreased the power of "program specialists," who were presumably strongly committed to "their" social and redistributional programs at the expense of "generalists" or elected officials more concerned with taxpayer pressure. Further, the distribution of block grants to states and the addition of allocation formulae in several important programs appeared directly aimed at what the administration perceived to be the undue advantages of large cities in the past (Craig and Inman 1982; Dommel and Rich 1987; Gramlich 1985; Kincaid 1987; Wong and P. E. Peterson 1986). Fourth, the relative power of the states in fiscal federalism was increased (Nathan, Doolittle, and Associates 1987; Wright 1978), which might have been expected to bring less interest in helping central cities and the urban "needy." Taken together, all these factors suggest a considerable deterioration in the support available to the urban (or for that matter, rural) poor.

In this chapter, we examine the impact of cutbacks in intergovernmental aid on U.S. cities to test the conventional images outlined above. We seek both to measure how great the cutbacks were and to assess their consequences for the public policies of U.S. cities. In the first empirical section, we present data measuring trends in the amounts of intergovernmental transfers. In the second section, we then discuss the comparative burdens of the decline in state and federal aid to cities between 1980 and 1989. We ask which types of cities received the most aid in 1980 and which suffered the greatest relative losses during the "Reagan revolution." Finally, we examine the impact of falling intergovernmental aid flows on urban budgets as a central indication of the consequences of changing fiscal federalism.

Trends in Intergovernmental Revenues

The postwar era witnessed a revolution in fiscal federalism as intergovernmental transfers rose from fairly insignificant levels to major proportions of state and local spending by the late 1970s. Since then, intergovernmental aid flows have declined, exacerbating the

TABLE 6.1 Federal Transfers to State and Local Governments

	Current $ Bil	*Constant 1982 $ Bil*	*Per Cap 1982 $*	*Percent GNP*	*Percent Fed Bud*	*St-Loc Own Buds*
1929	$0.1	$0.6	$5	0.1%	4.5%	1.4%**
1939	$1.0	$7.2	$55	1.1%	11.1%	7.0%**
1949	$2.2	$9.2	$62	0.9%	5.4%	5.3%**
1955	$3.2	$12.7	$77	0.8%	4.7%	10.2%
1960	$7.0	$24.7	$136	1.4%	7.6%	14.5%
1965	$10.9	$35.4	$182	1.6%	9.2%	15.1%
1970	$24.1	$61.2	$299	2.4%	12.3%	19.0%
1976	$59.1	$96.2	$441	3.5%	15.9%	24.1%
1978	$77.9	$109.7	$492	3.6%	17.0%	26.5%
1980	$91.5	$105.9	$464	3.4%	15.5%	25.8%
1982	$88.2	$88.2	$379	2.8%	11.8%	21.6%
1984	$97.6	$90.2	$382	2.6%	11.5%	20.9%
1986	$112.4	$97.0	$401	2.7%	11.3%	20.5%
1988	$115.3	$95.1	$376	2.4%	10.8%	17.7%
1989	$122.0	$96.6	$380	2.4%	10.7%	17.3%
1990	$135.4	$100.5	—	2.7%	10.8%	19.4%
1991	$152.0	$108.9	—	3.1%	11.5%	20.5%
1992*	$182.2	—	—	3.2%	12.3%	—

SOURCE: Advisory Commission on Intergovernmental Relations (1988, pp. 22-25, 1990, pp. 42, 64, 1992, p. 60).
NOTE: *Estimated; **1 or 2 years different from stated date.

fiscal pressures on U.S. cities and local governments. Tables 6.1 and 6.2 chart these trends in gross amount of federal transfers and intergovernmental aid as a percentage of state and local budgets. All, unfortunately, have doleful implications for city governments.

The first table shows that federal transfers were minuscule before the Great Depression and rose to about 1 percent of GNP on the eve of World War II, confirming the image that the New Deal basically involved direct program delivery by the federal government in large part because state and local governments simply could not handle the massive problems thrown up by the Depression. By 1960, federal aid had not risen much in terms of its ratio to GNP (from 1.1 to 1.4 percent) and had actually fallen in terms of its share of the federal budget (from 11.1 to 7.6 percent).

During the next decade and a half, though, the Great Society programs and the Nixon administration's New Federalism produced skyrocketing federal aid flows, although they were aimed at somewhat different constituencies. President Johnson's Great Society programs in the mid- and late 1960s brought the first major surge in federal grants-in-aid, with urban areas and the poor being especially targeted by many of these programs. In contrast, President Nixon elaborated an alternative in his call for a New Federalism in the early 1970s aimed at devolving power from the federal level and reversing Johnson's liberal redistributive policies. A keystone in this initiative was the General Revenue Sharing program, which provided primarily unrestricted aid to a huge variety of state and local governments (Dommel 1981; G. E. Peterson 1984; Reagan 1971). Ironically, in comparison to the previous categorical programs that were explicitly designed to be more redistributive, the block grants and formula funding of Nixon's New Federalism actually probably promoted redistribution. The new procedures made it easier for poorer and smaller communities to gain access to federal aid as the competition for categorical grants had been dominated by sophisticated grantsmen working for the better endowed communities (Stein 1982).

This surge in federal aid is clear in Table 6.1. Federal transfers approximately tripled in current dollars and doubled in constant dollars between 1960 and 1970, and jumped again by 150 percent in current and 50 percent in constant dollars during 1970-1976. This trend continued, albeit at a somewhat reduced rate, in the first half of the Carter administration as federal aid increased approximately one third in current dollars and one seventh in constant dollars between 1976 and 1978, peaking in real value in the latter year at $109.7 billion in 1982 dollars. By 1978, federal aid to state and local governments had risen to 3.6 percent of GNP and 17 percent of the federal budget. Federal aid also increased substantially, from 10 to 25 percent of total state and local expenditures, making it a very significant source of support for these lower levels of government. However, with the growing economic crisis of the late 1970s, the Carter administration began to reduce intergovernmental aid as an anti-inflation measure. As a result, although expenditures for federal transfers grew by one seventh in current dollars during its last 2 years, this was a slight decrease in real spending; federal transfers as a percentage of the federal budget, GNP, and state-local outlays fell slightly as well.

The downturn that started at the end of the Carter administration continued apace under the first several years of President Reagan's

New Federalism. Reagan's New Federalism had several broad objectives: (a) reducing federal transfers as part of the goals of cutting the domestic budget in general and transferring responsibilities to lower levels of government in particular, (b) merging past programs into large "block grants" that gave more discretion to state and local governments, and (c) reducing the priority of social programs at all levels of government. Again, the aggregate data in Table 6.1 are in accord with conventional images of the early 1980s and indicate that Reagan's New Federalism was successful in several regards. Federal aid was cut significantly, at least in the short term (Seymour et al. 1982 provide a detailed list of program cuts), and states and cities began paying more of their own bills. In addition, these figures show the small rebound in federal aid beginning in 1984 that has received recent analytic attention (Nathan and Lago 1988; Rafuse 1987; Wolman 1988).

Funds for the federal aid in current dollars actually fell from $92 to $88 billion between 1980 and 1982 and were only 7 percent higher in 1984 than the 1980 level. This created a one sixth drop in "real" 1982 dollars from $106 billion to $88 billion. Correspondingly, federal aid fell to 2.8 percent of GNP, 11.8 percent of the federal budget, and 21.6 percent of state and local revenues from their own sources—that is, just above the 1970 level. Following the nadir of 1982, federal grants increased somewhat to $115 billion in current dollars and $95 billion in constant dollars in 1988 and then jumped significantly under the Bush administration to $152 and $129 in 1991. However, the ratios of federal transfers to gross national product, the federal budget, and state-local revenues—which measure the weight of transfers relative to the overall economy and government budgets—continued to fall under Reagan and remained significantly under 1980 levels at the end of the Bush era.

The data in Table 6.2 on the ratio of intergovernmental transfers to a political unit's own revenues provide another angle for assessing the impact of the dramatic postwar changes in intergovernmental aid flows on city budgets. Before 1970, state flows were much more important than federal ones for local governments; federal transfers were concentrated at the state, rather than the local level. For example, in 1970 state transfers equaled 33 percent of city local revenues (up significantly from 19 percent in 1955) and 67 percent of county local revenues, whereas the corresponding figures for federal aid were 7 percent and 2 percent. Federal transfers were also quite

TABLE 6.2 Intergovernmental Transfers as Percent of Own Revenues

	Federal Transfers			*State Transfers*	
	State	*City*	*County*	*City*	*County*
1955	20.9	1.9	1.1	19.4	59.9
1965	32.3	4.5	1.6	22.2	53.7
1970	33.5	7.1	2.3	33.0	67.4
1974	35.5	19.8	14.2	38.0	66.2
1978	37.0	25.8	19.2	36.5	61.1
1980	36.6	22.8	16.6	33.4	63.6
1982	32.1	18.4	11.1	31.7	59.2
1984	30.5	14.5	8.5	28.6	52.1
1986	31.4	11.6	6.8	29.0	50.8

SOURCE: Advisory Commission on Intergovernmental Affairs (1988, p. 81). ACIR discontinued this series after 1986.

significant for states (33.5 percent of their own revenues, up from 20.9 percent in 1955), in contrast to their minuscule impact on city and county budgets. Thus, the first decade of the spurt in federal transfers (1959-1969) was clearly targeted on the states.

The second decade of expanding federal aid witnessed a much different pattern, with local governments the major beneficiaries of federal largess, whereas state support generally stabilized. Between 1970 and 1978 the ratio of federal transfers to local governments' own revenues (note that these are considerably higher than would be their ratios to total revenues) approximately quadrupled from 7 to 26 percent for cities and from 2 to 19 percent for counties; the corresponding proportion for states only rose from 33.5 to 37 percent. The ratio of state transfers to local government funds remained fairly constant as well, rising from 33 to 36.5 percent for cities and falling from 67 to 61 percent for counties. Thus, the expansion of urban budgets and responsibilities during the late 1960s and 1970s was substantially financed by increased federal aid.

This considerable jump in dependence on federal financing, in turn, made U.S. cities vulnerable when the economic crisis of the late 1970s led to substantial cuts in intergovernmental aid. Federal transfers as a percentage of cities' own budgetary resources dropped dramatically from a high of 25.8 percent in 1978 to 11.6 percent in 1986 (the corresponding fall for counties was 19.2 to 6.8 percent); the decline in federal aid to states was much more gradual (from 37 to

TABLE 6.3 Correlations Between Different Types of per Capita Intergovernmental Transfers

	Total and Federal Transfers	*Total and State Transfers*	*Federal and State Transfers*
1980 Level	.73*	.76*	.24*
1980-85 Change	.50*	.58*	–.01
1985 Level	.60*	.81*	.17*
1985-1989 Change	.45*	.46*	–.01
1989 Level	.42*	.85*	.13*
1980-1989 Change	.40*	.61*	.02

SOURCE: U.S. Census fiscal data compiled as part of Fiscal Austerity and Urban Innovation Project.
NOTE: *Statistically significant at .01 level.

31.4 percent). In addition, the ratio of state transfers to cities' own revenues fell from 36.5 to 29 percent. Thus, at the same time that federal program cutbacks were augmenting the strains and demands on urban governments, intergovernmental (federal plus state) aid fell precipitously in just 6 years from 62.3 to 40.6 percent of the financial resources raised by the cities themselves. Thus, the federal cuts of Reagan's New Federalism clearly had a much larger relative impact on cities than states. These aggregate data on intergovernmental transfers imply that federal and state intergovernmental transfers to U.S. cities follow quite different logics. This can only be directly ascertained, however, by analyzing the actual amounts of federal and state aid going to each city. Thus, Table 6.3 presents correlations between these three types of aid, measured per capita, going to cities with populations greater than 25,000 for our three time points (1980, 1985, and 1989) and for the temporal changes between them (1980-1985, 1985-1989, and 1980-1989). Table 6.4 gives the correlation for each type of transfer among levels at the three time points and changes.

Table 6.3 confirms two general images about the nature of state and federal intergovernmental transfers during the 1980s. First, state transfers clearly became much more important relative to federal transfers during the 1980s. For example, state and federal per capita transfers had approximately equal correlations with total receipts in 1980 (approximately .75), but by 1989 state funds had a correlation twice as high as federal ones (.85 to .42) with the total amount of intergovernmental aid to a specific city. Second, federal and state intergovernmental aid were

TABLE 6.4 Correlations Between Levels and Changes in Specific Types of Intergovernmental Transfers

	Total	*Federal*	*State*
1980 Level With 1980-1985 Change	–.35*	–.39*	–.28*
1980-1985 Change With 1985 Level	.26*	.35*	.18*
1980 Level With 1985 Level	.83*	.70*	.90*
1985 Level With 1985-1989 Change	.01	.01	–.09*
1985-1989 Change With 1989 Level	.48*	.69*	.25*
1985 Level With 1989 Level	.87*	.62*	.93*
1980 Level With 1980-1989 Change	–.17*	–.15*	–.18*
1980-1989 Change With 1989 Level	.45*	.66*	.61*
1980 Level With 1989 Level	.77*	.50*	.88*

SOURCE: U.S. Census fiscal data compiled as part of Fiscal Austerity and Urban Innovation Project.
NOTE: *Statistically significant at the .01 level.

allocated on rather different bases. In 1980, there was only a moderate correlation of .24 between the state and federal transfers going to U.S. cities. During the 1980s, there was no association at all between changes in federal and state transfers, so that by the end of the decade the relationship between their absolute levels had become minuscule ($r = .13$).

The data on the relationships among the levels of each type of aid at the three time points and their amounts of temporal change in Table 6.4 show that as might be expected, the allocation of state transfers was much more stable than for federal aid in the sense that the relative ranking of aid recipients was maintained during the 1980s. For state aid, the correlation of .88 between the 1980 and 1989 levels indicates little change in state aid, but for federal aid the correlation is much lower at .50. Thus, although cities that were most advantaged by federal transfers in 1980 still received more aid than others in 1989, there was far more change in the relative rankings of cities on federal than state aid. A priori, this is consistent with the image that the federal government cut back on its progressive transfers, whereas no countervailing change in the nature of state aid occurred.

In addition, the changes in intergovernmental flows differed quite significantly between the first and second terms of the Reagan administration. Interestingly, the rate of change for 1980-1985 had a moderately negative correlation with the 1980 level of receipts for both federal ($r = -.39$) and state ($r = -.28$) transfers that of course had to create a similar one for total aid ($r = -.35$). On the other hand, the rate of change had moderate positive correlations in the .20 to .35 with the 1985 per capita level of all three flows as well. Thus, the cities that were most advantaged in terms of per capita intergovernmental aid in 1980 suffered the largest cutbacks during the next half-decade, whereas the cities that received the largest gains (or least losses) tended to have the highest support in 1985. The amount of this redistribution was limited, though, because per capita receipts in 1980 and 1984 remained highly correlated (with *r*s between .70 and .90). For 1985-1989, a somewhat similar pattern occurred, but in a much more attenuated fashion. Rather than a negative association with 1985, change during 1985-1989 was uncorrelated with the base year, and its association with the resulting 1989 levels of intergovernmental aid was significantly higher than for change during the preceding half decade.

The somewhat redistributional nature of changes in intergovernmental flows during Reagan's first term, further, is consistent with Robert Stein's (1981) insightful demonstration that the cities that had the longest involvement with federal programs received the most aid, participated in the widest variety of programs (particularly categorical grants), and tended to have the greatest capacity for grantsmanship rather than the greatest need for grants. Thus, these cities were hurt the most by disproportionately large cuts in the categorical program, raising the possibility that Reagan's New Federalism (like Nixon's before it) might have unwittingly undercut its ostensible attack on the progressiveness of federal aid. The facts that this inverse relationship was stronger for federal than for state flows and that (as observed above) there was almost no association between changes in federal and state aid might be explained, hence, by the latter's greater nondiscretionary character, which limited the ability of even the most capable city officials in manipulating it.

Perceptions of city officials also dovetailed with the budgetary figures just described. The city CEOs questioned in the FAUI study indicated that the loss of federal and state revenues had created major problems for their governments, as indicated in Table 6.5. More than half said that the loss of state revenue had been important for their city; and 84 percent attributed at least some importance to the loss of state funds. The corresponding figures for loss of federal revenues were not quite as high (40 and 83 percent, respectively), but still

TABLE 6.5 Importance of Loss of Intergovernmental Funds

	State Revenue		*Federal Revenue*	
	N	*Percent*	*N*	*Percent*
Least importance	75	15.9	80	17.1
Some importance	140	29.6	202	43.1
Very important	136	28.8	148	31.6
Extremely important	122	25.8	39	8.3
TOTAL	473		469	

SOURCE: U.S. Fiscal Austerity and Urban Innovation Project survey of Chief Administrative Officers.

reflected the imposition of considerable fiscal stress. These perceptions of city officials might seem at variance with the previously discussed budget data that showed the drop in federal aid to be much more precipitous than for state aid. However, the absolute impact of state aid was greater because it constituted a much higher percentage of city budgets.

In sum, both budgetary data and the perceptions of city officials indicate that the fiscal stress facing U.S. cities as a result of regional and national economic crises in the late 1970s and early 1980s was significantly exacerbated by the loss of intergovernmental aid. Declining federal aid forced most cities to rely more on their own narrower resource bases and increased the role of states in transfers to cities and other local governments.

These trends imply a more regressive provision of urban public services for a variety of reasons. First, city revenue bases are generally narrower, less flexible, and less progressive than state and especially federal ones. Second, state replacements of federal cuts appeared to be much more oriented toward developmental and infrastructure projects than toward helping larger and poorer cities (Nathan and Doolittle 1985). Finally, poorer cities with higher fiscal strain were much less able than their better endowed counterparts to find local revenues to replace cuts in intergovernmental transfers (Wolman 1988). In the next section, we evaluate the extent to which this regressive potential was realized by empirically addressing two important questions: What types of cities were especially hurt by the changing nature of fiscal federalism? What effects, if any, did changes in federal and state aid flows have on budgetary policy?

The Causes and Consequences of Intergovernmental Flows to U.S. Cities

An important question for both theoretical and practical reasons concerns the characteristics of winners and losers among U.S. cities in the reduction and redistribution of intergovernmental aid during the early 1980s. In this section, we examine the association between intergovernmental transfers to an urban area and a city's socioeconomic and political characteristics at three points in time (1980, 1985, and 1989) marking the beginning, middle, and end of the Reagan administration.

The direction and relative impact of the various independent variables should be theoretically significant. The association, if any, between intergovernmental aid and economic traits will establish whether the decrease in intragovernmental revenues during the 1980s had the hypothesized regressive effects. Further, determining whether cities' political characteristics had any independent effects on the receipt of intergovernmental aid will indicate whether the system had some slack that could be manipulated from below and/or whether politically motivated targeting determined aid flows.

The analysis includes four standard socioeconomic variables—income per capita, population size, percent minority population, and population growth during the decade preceding the year being analyzed. These items should indicate the extent to which intergovernmental flows were progressive in the sense of being directed toward "needy" cities. Thus, examining their impact on the aid to various cities provides a direct test of the "increasing regressiveness" hypothesis.

Two political dimensions are also included. The first is the "administrative sophistication" of a city, which has been shown to have had a significant effect on city retrenchment strategies (C. Clark and Walter 1986; Morgan and Pammer 1988). Cities with more sophisticated and professional administrations should normally have an advantage in competing for intergovernmental aid, and cities' planning capabilities have had a considerable impact on their abilities to apply and compete for grants (Stein 1979). This advantage might be offset by either of two factors, though. First, such administrators might be hesitant to make their cities overly dependent on external funding and/or object to the redistributional programs entailed by specific federal grants. Thus, they might be expected to exercise "fiscal selectivity" in the pursuit of federal grants (T. N. Clark 1983;

T. N. Clark and Ferguson 1983; Saltzstein 1977). Second, some aid programs may be primarily dependent on political contracts and criteria rather than professional sophistication in preparing grant applications. For example, political contracts and lobbying clearly play a central role in the allocative decisions of many federal programs (Agnanoson 1982; Arnold 1979; Ferejohn 1974; Haider 1974).

Two empirical indicators of the administrative sophistication of a city were used here: (a) whether the city had a city manager as opposed to the more "political" mayor form of government and (b) an index of "fiscal management capability" based on how extensively a city used the fiscal monitoring and forecasting procedures reported in the FAUI survey (Appleton and T. N. Clark 1989). The former (but not the latter) indicator, it should be noted, also provides an inverse measure of political contacts and abilities that mayors would be expected to possess to a greater extent than managers (Haider 1974). For example, P. E. Peterson et al. (1986) found a major difference between the manner in which "politicized" and "professional" administrators reacted to federal programs.

The political culture and leadership style of a city have also had a considerable impact on fiscal retrenchment styles and reaction to cutbacks in intergovernmental aid (Appleton and T. N. Clark 1989; T. N. Clark 1983; T. N. Clark, Burg, and Landa 1984; T. N. Clark and Ferguson 1983; C. Clark and Walter 1988; Mollenkopf 1983; Wong and P. E. Peterson 1986; Wright 1978). Thus, indices derived from the FAUI survey measuring three of the urban cultures posited by Terry Clark and Lorna Ferguson (1983) were included in the analysis because they reflect leadership political cultures that might be expected to adopt an aggressive position toward garnering intergovernmental aid—Democrats, for their fiscal liberalism; Ethnic Politicians because of their association with high-cost patronage politics; and New Fiscal Populists (NFPs) because of their activism in confronting budget problems despite their fiscal conservatism.

To assess what types of cities were benefited or hurt by each of the intergovernmental transfer programs, we developed multiple regression models explaining the receipt of intergovernmental revenues by cities' socioeconomic and political characteristics. Initially, the four socioeconomic and five political indicators defined above were entered into three separate regression equations—one for each per capita level of aid received at three points in time: 1980, 1985, and 1989. Only the independent variables significant at the .05 level were

TABLE 6.6 Determinants of Federal Transfers per Capita

	1980	*1985*	*1989*
MULTIPLE *R*	.69**	.65**	.50**
Betas			
Income per Capita	−.52**	−.50**	−.24**
Percent Minority Pop	.14*	.12*	.25**
Population Size	.15**	.17**	.12*
Population Growth	−.23**	−.16**	−.14*
NFP Pol Cul	—	—	—
Democrat Pol Cul	—	—	—
Ethnic Pol Cul	—	—	—
Manager	—	—	—
Fiscal Mangmt Index	.12**	.14**	.11*

SOURCE: U.S. Fiscal Austerity and Urban Innovation Project.
NOTE: *Statistically significant at .05 level; **statistically significant at .01 level.

included in the final regression equations reported below to make the visual presentation clearer. The data on intergovernmental transfers and income per capita were subjected to log transformations.

Table 6.6 presents these regression results for federal transfers. In combination, the independent variables were quite important in structuring the amount of federal grants going to individual cities. The Multiple *R*s for both 1980 and 1985 were almost .70. However, the *R* for 1989 dropped somewhat to .54, suggesting a growing importance for distributive criteria that are neither tied to criteria of need nor subject to manipulation from below.

One of the most striking features of Table 6.6 is that a city's socioeconomic characteristics had a strong impact on the level of intergovernmental aid it received, but political effects were quite marginal. This is consistent with Rich's (1989) conclusion that the structure of federal programs rather than individual political relationships and lobbying are most important in determining how program funds are actually allocated. Because many federal programs are ostensibly tied to objective criteria, they would be expected to distribute intergovernmental transfers according to such factors as city population size and affluence. The influences on federal transfers were quite similar in 1980 and 1985 and certainly redistributive in the sense that income per capita had by far the greatest impact, as indicated by having betas twice as high as any other explanatory variable with poorer cities receiving the greatest aid (beta = about –.50 for both years). In addition, there was a moderate tendency for bigger cities and cities with large minority populations to receive more federal aid in both years—also consistent with a redistributive emphasis. Population growth exercised a moderate influence on federal transfers in both years with faster growing cities receiving less aid (beta = about –.20). This has countervailing implications for progressiveness. On the one hand, rapid growth brings the need for expanding costly public services (such as sanitation, schools, and streets); on the other, population growth is usually associated with prosperity. Finally, cities with sophisticated fiscal management systems garnered a little more federal support than those without these capabilities during the first half of the 1980s, implying continuation of a moderate but significant role for grantsmanship.

By 1989, in contrast, the nature of per capita federal transfers had changed significantly, but in a very surprising way. A considerable change occurred in the relationship between minority populations and federal aid; however, it was in the opposite direction of what would have been expected for the height of Reaganomics! Cities with large minority populations saw their federal aid increased substantially, so that the beta for this variable doubled to .25. Conversely, the beta for income per capita was halved to –.24, and effects of the other three significant independent variables stayed about the same. Thus, the federal aid was about as progressive in 1989 as it had been in 1980 and 1985. The only difference was that in the former year minority population and income per capita had about equal weights in attracting federal funds, whereas in the latter two affluence by itself was the decisive factor. This continuing progressiveness, therefore, supports the findings of Morgan and Shih (1991) that significant federal targeting of "needy" cities continued during the Reagan administration.

Federal transfers appear quite progressive, therefore; and the conservative Reagan administration surprisingly did not have much of an impact on this, although it certainly cut back federal aid as a percentage of city budgets. This challenges several previous conclusions that federal aid has not generally been targeted toward needy communities (Agnanoson 1980; Copeland and Meier 1984; Cucitti 1978; Dye and Hurley 1978; Gist and Hill 1981; Saltzstein 1977; Stein 1982). Almost all these studies, however, were based on fairly small samples in geographic area, community size, or program coverage.

TABLE 6.7 Determinants of State Transfers per Capita

	1980	*1985*	*1989*
MULTIPLE *R*	.39**	.35**	.33**
Betas			
Income per Capita	—	—	—
Percent Minority Pop	–.17**	–.14*	–.18**
Population Size	.17**	.19**	.19**
Population Growth	–.18**	—	—
NFP Pol Cul	—	—	—
Democrat Pol Cul	—	—	—
Ethnic Pol Cul	—	—	—
Manager	–.24**	–.26**	–.22**
Fiscal Mangmt Index	—	—	—

SOURCE: U.S. Fiscal Austerity and Urban Innovation Project.
NOTE: *Statistically significant at .05 level; **statistically significant at .01 level.

Thus, our results support several more recent and more comprehensive analyses that have been concluded that larger and poorer cities did receive more federal aid during the 1980s (Morgan and Shih 1991; Stein and Hamm 1987).

State transfers would be expected to differ considerably from federal ones; and Table 6.7 shows that this is indeed the case. First, the levels of state flows in all 3 years were much less structured than federal ones in that our socioeconomic and political factors had far less impact on them: all the Multiple *R*s were in the moderate range of .33 to .39. This is consistent with the observation that much of state aid is distributed according to automatic but untargeted allocation formulae (e.g., returning sales tax revenues to jurisdictions in which they were collected). Second, despite this, political factors were much more important at the state level than for the distribution of federal flows on which they had a relatively marginal impact, confirming Cole's (1990) argument that federal cutbacks had forced cities to reestablish their contacts with state agencies and officials.

The strength of the political variables in shaping patterns of state intergovernmental transfers represents a sharp contrast to the data on federal aid. For all 3 years (1980 1985, and 1989), cities with mayors did substantially better than those with managers in attracting state aid (beta = about –.25 for each year); in fact, this was the most important explanatory factor in all three regressions. Our results substantiate the theory that mayors could benefit from the presumed links with state officials in ways that "nonpolitical" managers could not and imply that state transfers possess a surprisingly strong political character.

A second strong contrast to the pattern of federal flows was that state transfers lacked a redistributional format and could even be considered somewhat regressive. Income per capita was unrelated to state aid, but cities with large minority populations were actually disadvantaged to a moderate extent by state allocations (beta = about –.15 for all 3 years); the balance seems to be slightly against cities exhibiting the normal signs of socioeconomic need. Finally in terms of the socioeconomic context, cities that were growing rapidly received less state funding than others (beta = –.18) in 1980, suggesting that a time lag occurs between increase in population and the commercial development that generates much of the tax money on which state transfers are based (C. Clark and J. Clark 1986 provide a detailed state case study showing this relationship). However, this evident bias had vanished by 1985.

These results can be explained by the nature of state intergovernmental aid. Many of these transfers are simply the "flow-through" distribution of revenues collected in a locality (e.g., sales taxes) that presumably benefit richer areas; and discretionary state aid funds, including the ones used in the early 1980s to "replace" federal cuts, tended to emphasize development rather than welfare or the needs of large cities (Nathan and Doolittle 1985; G. E. Peterson 1984). Thus, the lack of progressiveness is easily explicable. Interestingly, although these analysts were certainly correct in noting the much less progressive nature of state compared to previous federal allocations, the states' neutrality regarding the size and affluence of cities receiving their transfers paralleled their pre-Reagan predispositions.

These findings about the nature of state transfers, however, are somewhat at variance with several earlier studies that concluded that states did target their transfers toward needy communities to a significant extent (Dye and Hurley 1978; Hawkins and Smith 1987; Pelissero 1984, 1985), although some of these studies just examined specific programs rather than the total net state flows. In contrast, several more comprehensive and sophisticated analyses have indicated that only a few states actively target "needy" communities in their distribution of intergovernmental aid (Morgan and Shih 1991; Stein 1982; Stein and Hamm 1987). Thus, the conclusion that state

TABLE 6.8 Determinants of Total Transfers per Capita

	1980	*1985*	*1989*
MULTIPLE *R*	.53**	.50**	.37**
Betas			
Income per Capita	−.26**	−.21**	—
Percent Minority Pop	—	—	—
Population Size	.15**	.22**	.20**
Population Growth	−.28**	−.18**	—
NFP Pol Cul	—	—	—
Democrat Pol Cul	—	—	—
Ethnic Pol Cul	—	—	—
Manager	−.15**	−.22**	−.27**
Fiscal Mangmt Index	—	.11*	.14*

SOURCE: U.S. Fiscal Austerity and Urban Innovation Project.
NOTE: *Statistically significant at .05 level; **statistically significant at .01 level.

transfers in general were not progressive appears sound. In fact, the aggregate data analyzed here suggest another reason for this. Although some programs in some states may be well targeted and highly redistributive, they are probably offset in total state transfers by "flow-through" returns of taxes collected in specific communities that are almost inevitably regressive.

Further, discretionary state transfers evidently had an important political component in that more politically attuned mayors were able to outcompete administratively oriented managers.

The presumed political contacts of mayors and activist city officials, therefore, worked at the state level, but not in the pursuit of federal funds, for which the distribution is certainly subject to political influence. This implies that other actors (e.g., congressional representatives and interest groups) were more important in "pork barrel politics" at the federal level and that the prominent success of a few big-city mayors (e.g., Richard Daley of Chicago) in tapping the federal till cannot be generalized to most municipal leaders (Haider 1974).

Total intergovernmental transfers represent the combined impact of state and federal flows, so that some of their countervailing effects would be expected to cancel each other out. Table 6.8 presents the regression results for all intergovernmental transfers (the role of local government transfers to cities in total flows is minuscule). The Multiple *R*s for 1980 and 1985 are at the moderate level of about .50, but that for 1989 drops to .37 as the more unstructured state transfers composed a higher percentage of the total mix of monies to cities.

The growing weight of state aid also caused a major shift in the determinants of city receipts of intergovernmental aid between 1980 and 1989. In 1980, socioeconomic effects were most important and operated in a progressive manner in the sense that poorer (beta = –.26) and larger (beta = .21) cities were advantaged by the system. Cities with low or negative population growth also received high levels of intergovernmental transfers (beta = –.28), which as explained above has ambiguous implications for progressiveness. Although the socioeconomic factors were most important, some political effects could be discerned as well. In particular, there was a moderate tendency for cities with mayors to do well in the allocation of intergovernmental aid (beta = –.15) because of their advantage in tapping state funds.

By 1989, this pattern of relationships had changed considerably (1985 was fairly similar to 1980). Political factors had become more important in shaping intergovernmental flows than socioeconomic ones. The strongest explanatory factor in 1989 had become the presumed political influence at the state level exercised by mayors (beta = –.27) as opposed to less well connected city managers. Further, there was a slight tendency for cities with high management capabilities (beta = .14) to reel in higher levels of support, suggesting at least a little continued efficacy for grantsmanship. Thus, the two indicators of administrative sophistication had surprisingly countervailing effects because of their different relationships with state and federal transfers. In short, by the end of the Reagan era the growing stress on city finances had seemingly forced cities to bring all their political capabilities into play in attempts to continue old or find new sources of external funding, primarily at the state level.

In terms of the impact of socioeconomic factors, the progressiveness of 1980 had clearly been lost by 1989. The advantage that poorer cities had in 1980 and 1985 disappeared as the ability of richer cities to make the largest gains (or suffer the smallest losses) at the state level came to cancel out the continued progressiveness in federal allocations. Moreover, this was not an offset, as it had been for just federal funds, by growing generosity toward communities with large minorities, because the state shift in funds away from these cities offset their gains from federal transfers. The only significant associa-

TABLE 6.9 Determinants of City Expenditures per Capita

	1980	*1985*	*1989*
MULTIPLE *R*	.75**	.71**	.68**
Betas			
Intergov transfers p.c.	.69**	.54**	.49**
Income per Capita	.12*	.11*	—
Percent Minority Pop	—	—	—
Population Size	.22**	.40**	.40**
Population Growth	—	—	—
NFP Pol Cul	—	—	—
Democrat Pol Cul	—	—	—
Ethnic Pol Cul	—	—	—
Manager	—	.12*	.15**
Fiscal Mangmt Index	—	—	—

SOURCE: U.S. Fiscal Austerity and Urban Innovation Project.
NOTE: *Statistically significant at .05 level; **statistically significant at .01 level.

tion between socioeconomic traits and total intergovernmental aid in 1989 was that large cities remained somewhat advantaged.

The adage "the proof of the pudding is in the eating" certainly has some merit. Thus, we also examined the impact of intergovernmental transfers on the spending levels of U.S. cities during the 1980s. Given the significant shares of urban budgets that intergovernmental transfers constituted in 1980, the substantial changes in these flows during the 1980s might well have had a direct impact on city expenditures and public policies. If changing intergovernmental support were a major determinant of budgetary shifts, the declining progressiveness of external support would have especially hurt communities whose populations are conventionally considered in the greatest need of public services.

To test this scenario about the impact of intergovernmental transfers on urban budgets, we regressed city spending levels in 1980, 1985, and 1989 on total intergovernmental transfers plus the nine socioeconomic and political indicators used in the previous analyses. Results are in Table 6.9. The explanatory factors taken together had a substantial effect on city spending levels throughout the 1980s, as indicated by the three Multiple *R*s in the .70 to .75 range.

The beta (standardized regression) coefficients clearly indicate that intergovernmental support had a considerable impact on the budgetary fates of U.S. cities during the 1980s, but one that declined over time (as would be expected given the growing reliance on their own revenues that was forced on urban areas during the decade). In 1980, total intergovernmental aid had by far the greatest impact on city spending levels (beta = .69), with slight tendencies for larger and more affluent cities to spend more. In 1985 and 1989, in contrast, intergovernmental receipts were still the most important independent variables (beta = about .50), but population size had moved up into a close second (beta = .40). This suggests that political and social factors (union power and the needs of the poor) kept spending levels up in large central cities, whereas smaller and more affluent suburbs were able to cut costs. Affluence per se was surprisingly unimportant (a slight influence in 1985 and none at all in 1989), suggesting the overriding importance of the political and social factors alluded to above. Finally, in both 1985 and 1989, cities with a council-manager form of government had somewhat higher spending than those with a mayor-council form, despite the latter's greater success in capturing state aid.

On balance, therefore, the changing pattern of fiscal federalism induced by the Reagan administration's New Federalism reduced the budgetary resources of the cities with the greatest social needs to a significant extent. The progressiveness in the overall nature of the intergovernmental transfer system waned despite the continued targeting, intended or not, of federal flows toward cities with large numbers of low-income and minority citizens. In turn, the very strong impact of intergovernmental aid on city expenditures, coupled with the biases at the local level against redistributive policies, almost certainly had a regressive effect. The magnitude of this change should not be overstated, however, as the total aid could not really be said to have become regressive in the sense that more affluent communities became clear-cut winners.

Cuts, Cultures, and City Limits

From a peak in the late 1970s, intergovernmental aid to cities in the United States dropped dramatically in terms of real dollars and as a proportion of both donors' and recipients' budgets. The cutbacks in federal transfers were also proportionately greater than those in state flows; the Reagan revolution made the states more predominant.

This constituted a grave challenge to municipal leaders both because the cities' own resource bases saw increased stress and because aid flows had played a major role in expanding public services in the 1960s and 1970s. In this chapter, we have used aggregate data on intergovernmental transfers to evaluate the paradoxical conventional image that these cutbacks were substantial but had only a limited impact on the operations of city governments.

It was widely expected that Reagan's New Federalism would have two major deleterious results. First, the loss of funds would quickly create financial disaster for many cities already under considerable fiscal stress. Second, the changes of the early 1980s held the potential for growing regressiveness in the system of fiscal federalism. The Reagan administration would have been expected to direct federal aid away from the larger and poorer cities; the growing share of state transfers in the total mix should have produced less progressive results; and growing reliance on their own funds should have decreased the priority of redistributive policies within the cities themselves. The dire predictions about financial crisis did not occur in the short run because of a combination of factors: cities adopted austerity measures and found new revenue sources; intergovernmental transfers picked up again after the sharp cuts of the early 1980s; and perhaps more important, the strong economic recovery of the 1980s filled city coffers. However, in the early 1990s a long recession and growing financial problems at all levels of government raised the specter of urban fiscal collapse again in many cities.

The results concerning the Reagan revolution's allegedly regressive nature are more complex. On the one hand, despite the expectations of both the Reaganauts and their sharpest critics, little actual change occurred in the direct overall progressive impact of federal aid throughout the 1980s; and the overall system of intergovernmental transfers remained significantly progressive during the first Reagan administration when the cuts in federal aid actually occurred. On the other hand, there was no change in the lack of progressiveness (and even regressive disadvantaging of cities with large minority populations) in state transfers. Thus, the growing importance of state aid ultimately destroyed the progressive nature of the overall system of fiscal federalism during the second Reagan administration.

The growing importance of state transfers in the overall mix of intergovernmental aid had a second important consequence—the increased influence of political factors on the receipt of intergovernmental transfers. This occurred in a three-step process. First, federal cuts forced cities to turn to the states, where city-level leaders had more potential direct contacts and influence than in distant Washington. Second, this shift resulted in the political variables supplanting the economic ones as the primary determinants of the allocation of state aid. Finally, the growth of state relative to federal transfers meant that their politically driven nature would ultimately become the most important factor in structuring the amount of external aid flowing to individual cities.

These political dynamics, further, might explain why Reagan's New Federalism seemed to match the failure that Stein (1982) noted in Nixon's New Federalism to shift intergovernmental transfers away from larger and poorer cities. The major reason for this is that fiscal federalism and city budget making occur within the socioeconomic and political matrix adumbrated by T. N. Clark and Ferguson (1983), P. E. Peterson (1981), and Wong (1988). The "limits" of a city's socioeconomic and political environment, therefore, interact to produce continuity and dampen the "revolutions" of new presidents.

From another perspective, however, the results seem far less sanguine in that "city limits" may well overwhelm "city cultures." The Peterson model implies that redistributive services for the poor have a low priority for cities and are generally stimulated by external mandates and aid, so that the federal cutbacks should have hit the urban poor especially hard. Moreover, none of the types of leaders adumbrated by T. N. Clark and Ferguson (1983) that were studied here—activist NFPs, Liberal Democrats and patronage-oriented Ethnic Politicians—seemed to be able to affect external aid although they had some capabilities (NFPs) or motivations (Democrats/Ethnic Politicians) for so doing, indicating "political limits" despite the success of mayors in finding some state support. Thus, city solvency in the 1980s might well have been expected to have been bought at the price of ignoring traditional social needs. Detailed empirical analysis of New Federalism during the early 1980s substantiated these predictions and researchers concluded that individuals rather than city governments suffered the most from the federal cuts (Nathan et al. 1982; Nathan and Associates 1987; Nathan and Lago 1988).

Given the persuasive argument that city limits preclude much local emphasis on redistributive programs, any new attempt for a skirmish, much less a war, on poverty almost inevitably must come from the federal government. It could certainly take very varied forms—from direct income maintenance transfers for individuals to a potpourri of narrow federal-state-local cooperative programs similar to the Great Society approach (ironically, P. E. Peterson et al. 1986 concluded that many of the Great Society's discredited redistributional programs were just beginning to work well due to the learning of both federal and local administrations when the Reagan era began). Thus, fiscal federalism should remain important in the discussion of the fundamental public policy alternatives facing the United States.

Our findings, to sum, certainly indicate that U.S. cities are highly dependent on the complex system of fiscal federalism that has evolved since the New Deal. This in turn implies that intergovernmental flows are an important field of study for scholars interested in understanding the dynamics of urban policy making and city money. More important, the "benign neglect" of intergovernmental transfers by policy makers needs to be overcome to construct a system more consistent with providing better public services to and improving the quality of life in U.S. urban centers.

References

Advisory Commission on Intergovernmental Relations. 1988. *Significant Features of Fiscal Federalism, 1988 Edition,* Vol. 2. Washington. DC: U.S. Government Printing Office.

———. 1990. *Significant Features of Fiscal Federalism, 1990 Edition,* Vol. 2. Washington, DC: U.S. Government Printing Office.

———. 1992. *Significant Features of Fiscal Federalism, 1992 Edition,* Vol. 2. Washington, DC: U.S. Government Printing Office.

Agnanoson, J. Theodore. 1980. "The Politics of the Distribution of Federal Grants: The Case of the Economic Development Administration." Pp. 61-91 in *Political Benefits: Empirical Studies of American Public Programs,* edited by Barry Rundquist. Lexington, MA: Lexington.

———. 1982. "Federal Grant Agencies and Congressional Election Campaigns." *American Journal of Political Science* 26, 3(August):547-61.

Appleton, Lynn M. and Terry Nichols Clark. 1989. "Austerity and Innovation in American Cities." Pp. 31-68 in *Urban Innovation and Autonomy: Political Implications of Policy Change,* edited by Susan E. Clarke. Newbury Park, CA: Sage.

Arnold, R. Douglas. 1979. *Congress and the Bureaucracy: A Theory of Influence.* New Haven, CT: Yale University Press.

Bueno, Edgar. 1992. "Local Economic Development as a Response to Federal Retrenchment." Paper presented as B.A. thesis, University of Chicago.

Clark, Cal and Janet Clark. 1986. "Federal Aid to Rural Western Governments: The Case of Wyoming Counties and General Revenue Sharing." Paper presented at the Annual Meeting of the Western Political Science Association, Eugene, Oregon.

——— and Oliver Walter. 1986. "City Fiscal Strategies: Economic and Political Determinants." Pp. 89-113 in *Research in Urban Policy: Managing Cities*, Vol. 2, Part B, edited by Terry Nichols Clark. Greenwich, CT: JAI.

——— and Oliver Walter. 1988. "Political Cultures, Administrative Styles, and Fiscal Austerity Strategies at Different Levels of Fiscal Strain." Paper presented at the Annual Meeting of the Midwest Political Science Association, Chicago.

Clark, Terry Nichols. 1983. "Local Fiscal Dynamics Under Old and New Federalisms." *Urban Affairs Quarterly* 19,1 (September):55-74.

———, Margaret M. Burg, and Martha Diaz Villegas de Landa. 1984. "Urban Political Cultures and Fiscal Austerity Strategies." Paper presented at the Annual Meeting of the American Political Science Association, Washington.

——— and Lorna Crowley Ferguson. 1983. *City Money: Political Processes, Fiscal Strain, and Retrenchment.* New York: Columbia University Press.

Clarke, Susan E. and Gary Gaile. 1991. "The Next Wave: Post-Federal Local Economic Development Strategies." Paper presented at the Summer Conference on Leadership and Economic Development, University of Chicago.

Cole, Richard L. 1990. "America's Ailing Cities and the 1980s: The Legacy of the Reagan Years." *Journal of Urban Affairs* 12, 4(December):345-60.

Copeland, Gary M. and Kenneth J. Meier. 1984. "Pass the Biscuits, Pappy: Congressional Decision-Making and Federal Grants." *American Politics Quarterly* 12, 1(January):3-21.

Craig, Steven G. and Robert P. Inman. 1982. "Federal Aid and Public Education: An Empirical Look at the New Fiscal Federalism." *Review of Economics and Statistics* 64, 4(November):541-52.

Cucitti, Peggy. 1978. *The Role of Equalization in Federal Grants.* Washington, DC: U.S. Government Printing Office.

Danziger, Sheldon and Daniel Feaster. 1985. "Income Transfers and Poverty in the 1980s." Pp. 89-117 in *American Domestic Priorities: An Economic Appraisal*, edited by John M. Quigley and Daniel L. Rubinfeld. Berkeley: University of California Press.

Dommel, Paul R. 1981. "Trends in Intergovernmental Relations: Getting Less, but Enjoying It More. Maybe." Pp. 91-108 in *Financing State and Local Governments in the 1980s: Issues and Trends*, edited by Norman Walzer and David L. Chicoine. Cambridge, MA: Oelgeschlager, Gunn, and Hain.

——— and Michael J. Rich. 1987. "The Rich Get Richer: The Attenuation of the Targeting Effects of the Community Development Block Grant Program." *Urban Affairs Quarterly* 22, 4(June):552-79.

Dye, Thomas R. and Thomas L. Hurley. 1978. "The Responsiveness of Federal and State Government to Urban Problems." *Journal of Politics* 40,1 (February):196-207.

Ferejohn, John A. 1974. *Pork Barrel Politics: Rivers and Harbors Legislation, 1947-1968.* Stanford, CA: Stanford University Press.

Gist, John R. and R. Carter Hill. 1981. "The Economics of Choice in the Allocation of Federal Grants: An Empirical Test." *Public Choice* 36(1):63-73.

Gramlich, Edward M. 1985. "Reforming U.S. Federal Fiscal Arrangements." Pp. 34-69 in *American Domestic Priorities: An Economic Appraisal*, edited by John M. Quigley and Daniel L. Rubinfeld. Berkeley: University of California Press.

Haider, Donald H. 1974. *When Governments Come to Washington: Governors, Mayors, and Intergovernmental Lobbying*. New York: Free Press.
Hawkins, Brett W. and Gregg W. Smith. 1987. "Conditions of State Aid to Distressed Communities." *Urban Affairs Quarterly* 23, 1(September):126-39.
Heilman, John and Gerald W. Johnson. 1992. *The Politics and Economics of Privatization: The Case of Wastewater Treatment*. Tuscaloosa, AL: University of Alabama Press.
Kincaid, John. 1987. "The State of American Federalism—1986." *Publius* 17, 3(Summer):1-33.
Marando, Vincent. 1990. "General Revenue Sharing: Termination and City Responses." *State and Local Government Review* 22, 3(Fall):98-107.
Mollenkopf, John H. 1983. *The Contested City*. Princeton, NJ: Princeton University Press.
Morgan, David R. and William J. Pammer, Jr. 1988. "Coping With Fiscal Stress: Predicting the Use of Financial Management Practices Among U.S. Cities." *Urban Affairs Quarterly* 24, 1(September):69-86.
——— and Mei-Chiang Shih. 1991. "Targeting State and Federal Aid to City Needs." *State and Local Government Review* 23, 2(Spring):60-68.
Nathan, Richard P., Phillip M. Dearborn, Clifford A. Goldman, and Associates. 1982. "Initial Effects of the Fiscal Year 1982 Reductions in Federal Domestic Spending." Pp. 315-77 in *Reductions in U.S. Domestic Spending: How They Affect State and Local Governments*, edited by John William Ellwood. New Brunswick, NJ: Transaction Books.
Nathan, Richard P. and Fred C. Doolittle. 1985. "Federal Grants: Giving and Taking Away." *Political Science Quarterly* 100, 1(September):53-74.
——— and Associates. 1987. *The Consequences of Cuts: The Effect of the Reagan Domestic Program on State and Local Government*. Princeton, NJ: Princeton Urban and Regional Resource Center.
———. 1987. *Reagan and the States*. Princeton, NJ: Princeton University Press.
Nathan, Richard P. and John R. Lago. 1988. "Intergovernmental Relations in the Reagan Era." *Public Budgeting and Finance* (Fall):15-29.
Pelissero, John P. 1984. "State Aid and City Needs: An Examination of Residual Aid to Large Cities." *Journal of Politics* 46, 3(August):916-35.
———. 1985. "Welfare and Education Aid to Cities: An Analysis of State Responsiveness to Needs." *Social Science Quarterly* 66, 2(June):444-52.
Peterson, George E. 1984. "Federalism and the States: An Experiment in Decentralization." Pp. 217-59 in *The Reagan Record: An Assessment of America's Changing Domestic Priorities*, edited by John L. Palmer and Isabel Sawhill. Cambridge, MA: Ballinger.
Peterson, Paul E. 1981. *City Limits*. Chicago: University of Chicago Press.
———, Barry G. Rabe, and Kenneth K. Wong. 1986. *When Federalism Works*. Washington, DC: Brookings Institution.
Rafuse, Robert W., Jr. 1987. "Fiscal Federalism in 1986: The Spotlight Continues to Swing Toward the States and Local Governments." *Publius* 17,3(Summer):35-53.
Reagan, Michael D. 1971. *The New Federalism*. New York: Oxford University Press.
Rich, Michael J. 1989. "Distributive Politics and the Allocation of Federal Grants." *American Political Science Review* 83, 1(March):193-213.
Saltzstein, Alan L. 1977. "Federal Categorical Aid to Cities: Who Needs It Versus Who Wants It." *Western Political Quarterly* 30, 3(September):377-83.

Seymour, Rita, Catherine Eschbach, John Gunther-Mohr, Charles Cameron, Dwight Dively, and John W. Ellwood. 1982. "Descriptions of Forty Major Reductions Affecting State and Local Governments." Pp. 133-311 in *Reductions in U.S. Domestic Spending: How They Affect State and Local Governments,* edited by John William Ellwood. New Brunswick, NJ: Transaction Books.

Stein, Robert M. 1979. "The Allocation of Federal Aid Monies: The Synthesis of Demand-Side and Supply-Side Explanations." *American Political Science Review* 75, 2(June):334-43.

———. 1982. "The Allocation of State Aid to Local Governments: An Examination of Interstate Variation." In *State and Local Roles in the Federal System.* Washington, DC: Advisory Commission on Intergovernmental Relations.

——— and Keith E. Hamm. 1987. "A Comparative Analysis of the Targeting Capacity of State and Federal Intergovernmental Aid Allocations: 1977, 1982." *Social Science Quarterly* 68, 3(September):447-65.

Wolman, Harold. 1986. "The Reagan Urban Policy and Its Impact." *Urban Affairs Quarterly* 21, 3(March):311-35.

———. 1988. "Fiscal Stress and Center-Local Relations: The Critical Role of Government Grants." Paper presented at the Annual Meeting of the American Political Science Association, Washington, DC.

Wong, Kenneth K. 1988. "Economic Constraint and Political Choice in Urban Policy-Making." *American Journal of Political Science* 32, 1(February):1-18.

——— and Paul E. Peterson. 1986. "Urban Response to Federal Program Flexibility: Politics of Community Development Black Grants." *Urban Affairs Quarterly* 21, 3(March):293-309.

Wright, Deil S. 1978. *Understanding Intergovernmental Relations: Public Policy and Participants' Perspectives in Local, State, and National Governments.* North Scituate, MA: Duxbury.

7

Contracting Out

A Solution With Limits

Rowan A. Miranda

NUMEROUS STUDIES COMPARING the efficiency of public and private service production conclude that service contracting can lead to lower costs (Ahlbrandt 1973; Bennett and Johnson 1979; Berenyi and Stevens 1988; Kemper and Quigley 1976; Savas 1987). Competition, scale economies, property rights, transaction costs, nonunion labor, and greater flexibility of the production function (or "input rigidities") are some of the reasons cited for the relative efficiency of private producers (Miranda 1994). Findings from these studies generally support the public choice view that competition undermines "Leviathan" (Bennett and Johnson 1979; Borcherding 1977). In *Reinventing Government*, Osborne and Gaebler (1992) call for the replacement of monopoly government with one that is competitive. Competitive governments typically separate the *steering* of service delivery (i.e., oversight and management) from *rowing* (i.e., the actual delivery of the service) (Ostrom, Tiebout, and Warren 1961).

For all the controversy it generates, and claims made on its behalf, considerable uncertainty persists about the policy effects of contracting out among both practitioners and scholars. First, most studies comparing relative efficiencies of public bureaus and private organizations do not address whether contracting reduces the "size" (e.g., expenditure and/or employment) of government. With few exceptions (Deacon 1979; Ferris 1988; Stein 1990; Miranda 1992c), the studies at best find one arrangement *can* produce services at a lower cost than another, but the studies do not link the superior private sector performance to reductions in *aggregate* spending or employment.

Second, the literature classifies a broad range of alternative service delivery arrangements (ASDAs) under the headings "public" or "private." What the discipline of public administration characterizes as the "blurring of sectors" (Bozeman 1987) is mostly ignored in the empirical literature on contracting. In deciding whether to contract out a service, municipal officials have to make choices concerning the organizational form of contracting. For example, human service programs can be contracted to a private for-profit corporation, a nonprofit organization (e.g., community or neighborhood group), or another governmental unit (e.g., county) (Ferris and Graddy 1986, 1988). Thus, the main question examined here is: *If there are expenditure and employment reductions from contracting some services, are they linked with some sectors more than others (i.e., private, nonprofit, or governmental)?*

In Section I, I review theoretical and empirical efforts examining the linkage between contracting out and municipal retrenchment. Section II contains the main hypotheses examined in this study and discusses the research design. Based on the 1982 International City Management Association (ICMA) Survey of Alternative Service Delivery Approaches, empirical findings from a sample of approximately 1,300 U.S. cities are discussed in Section III.

I. Contracting Out and Government Retrenchment

Two main theoretical frameworks—public choice and property rights—have guided empirical inquiry in studies comparing public and private production. Yet both theory and empirical research in these traditions have generally remained silent on the role of the "third sector" in service delivery.

Studies by Coase (1937, 1960), Alchian (1965), De Alessi (1969), and Alchian and Demsetz (1972) are illustrative of the *property rights approach* which posits that the most important difference between public and private firms is the nature of ownership. Private firms have a bottom line of profit, which is easy to monitor and evaluate, and dissatisfied shareholders can sell shares if profits drop. By contrast citizens as the "owners" of governments are not provided such a clear evaluation of performance, because governments pursue many goals, and few are precisely measured. Although citizens can move out of a town or protest an administrative action, these actions demand more from them than selling shares. Consequently, the property rights approach does not foster optimism about the prospects of public managers organizing factors of production to maximize the wealth of citizens. For our purposes, the main prediction from the property rights approach is that public sector production of the same service costs more than private sector production because the public sector lacks a mechanism that assigns ownership rights.

Public choice theory also predicts poorer government performance than the private sector, but does not base this expectation on the assignment of ownership rights. Instead, like neoclassical economics, public choice theorists emphasize the *lack of competition* facing public bureaus. The major theoretical issues related to service delivery are discussed by Niskanen (1971), whose model posits that budget maximizing "bureaucrats in public firms push their programs more strongly than would their private counterparts and join with those in the legislature who find such excess supplies congenial to their constituents' interests" (Borcherding, Burnaby, Pommerehne, and Schneider 1982, p. 137). What allows bureaucrats to be successful at program expansion is their monopoly of information vis-à-vis oversight authorities. In summary, the public choice approach not only predicts the outcome of the property rights approach (more *expensive* output) but also predicts a systematically *higher* level of output in the public sector.

Both the property rights and public choice theoretical frameworks have little to say about nonprofits. The emergence of the nonprofit sector in the United States has been carefully documented. Weisbrod (1988, p. 62) found that the number of organizations with tax-exempt status tripled in fewer than 20 years (1969-1985). Why might contracting with nonprofits be attractive?

The *contract failure theory* (Hansmann 1987, p. 29) suggests that in some instances "consumers feel unable to evaluate accurately the quantity or quality of a service a firm produces for them." Consequently, nonprofits appear attractive to consumers because of the "nondistribution constraint" that provides managers of nonprofits with fewer incentives to take advantage of their customers. Citing evidence from a study by Nelson and Krashinsky (1973), Hansmann (1987) states that:

> parents might wish to patronize a service provider in which they can place more trust than they can in a proprietary firm, which they might reasonably fear could take advantage of them by providing services of inferior quality. In such circumstances, a for-profit firm has both the incentives and the opportunity to take advantage of customers by providing less service to them than was promised and paid for. (p. 29)

The *subsidy theory* suggests that explicit and implicit subsidies, such as tax exemptions from federal, state, and local authorities, are largely responsible for the proliferation of nonprofit firms (Hansmann 1987, p. 33). Thus, observed differences in the cost structures of nonprofit firms must take into account subsidies that allow them to have an "edge" over the for-profit competition. Based on his earlier study of hospitals, Hansmann (1987) argues that "the availability of state, property, sales, and income tax exemptions has a significant effect in enhancing the market share of nonprofit firms vis-a-vis their proprietary competitors" (p. 33).

In comparison to public and private for-profit organizations, where might nonprofits lie with respect to cost efficiencies? On the one hand, competition in the nonprofit environment is closer to that expected in the private for-profit sector than in government. However, because assignment of property rights is inappropriate, this makes them quite similar to the public sector. The organizational form of nonprofits lies "between governments and markets," but does their relative efficiency do so also? (Ferris and Graddy 1991). Bozeman's (1987) *All Organizations Are Public* suggests that clear answers to these issues will be difficult to establish. Questions concerning the relative efficiency of nonprofits vis-à-vis government agencies and for-profit firms are thus a matter for further empirical inquiry.

These theories concerning the performance of alternative sectors have led to numerous efforts to examine cost differences between public and private production (see Donahue 1989; Miranda 1994; Savas 1987). In the case of refuse collection, the most sophisticated studies—those that control for scale, service levels, technology, and environmental factors—find that public production is approximately 35 percent more costly than contract collection, although a savings range from 14 to 124 percent is reported (Savas 1987). On the whole the studies suggest that *competition,* and not the *publicness* of the environment in which governmental service producers operate, is the principal determinant of efficiency.

Considerably fewer studies compare the relative efficiency of non-profit organizations to other sectors. Nevertheless, the evidence that does exist is mixed. For example, studies by Bays (1979), Wilson and Jadlow (1982), and Robinson and Luft (1985) find that for-profits are more cost efficient than nonprofits. These studies also show that public hospitals have lower costs than nonprofits. Other studies (Becker and Sloan 1985; Watt, Renn, Hahn, Derzon, and Schramm 1986) find no relationships between organizational form and service production costs. With the burgeoning role of the nonprofit sector in the U.S. economy (Salamon 1992; Weisbrod 1988; O'Neill 1989), it is important to examine the policy effects of contracting with this sector.

Studies comparing the costs of different organizational forms often assume rather than demonstrate a linkage between contracting and public expenditure. Just because a contractor produces a service at lower cost does not imply that these savings are automatically translated into reductions in government size. Indeed, an extension of the Niskanen model—the discretionary budget maximizing bureaucrat hypothesis (Migue and Belanger 1974)—implies that when public managers do generate savings from contracting, they face strong incentives to retain the savings within their bureaus (Stein 1990; Miranda 1992a, 1992b). Thus, at least one more step is needed to examine whether contracting out promotes municipal savings: establishing a linkage between contracting and reductions in aggregate expenditures and employment.

Several efforts have already been made to examine this linkage. Deacon's (1979) study of Lakewood Plan (i.e., contracting) cities in California finds expenditures to be approximately 20 percent lower than in non-Lakewood cities. Using larger samples of cities than Deacon, a series of studies based on the 1982 ICMA Survey of Alternative Service Delivery Approaches also show that contracting is associated with lower expenditures. Ferris (1988) finds that contracting has a statistically significant negative effect on expenditures and employment. Using a larger sample of cities and different models, Stein (1990) and Miranda (1992a) reach similar conclusions. Yet none of the studies attempted to link reductions in aggregate expenditures and employment to specific sectors contracted with.

II. Hypotheses and Measurement

Data on the municipal service contracting used in this study were obtained from the International City Management Association's 1982 Survey of Alternative Service Delivery Approaches. Approximately 3,130 cities were surveyed on the types of services they are responsible for providing and on the service delivery arrangements used to produce those services. This study examines all cities greater than 10,000 in population in the survey, with a sample size of approximately 1,330 cities.

The key measure used to examine the effects of contracting on aggregate expenditures and employment was the percentage of services that a city is responsible for that are contracted out (Table 7.1). Variables representing percentage of services contracted out by sector (e.g., percent of highway functions contracted to nonprofit organizations) were constructed by grouping individual services into specific agencies through an approach developed by Stein (1990). For example, individual services listed in the ICMA questionnaire, such as programs for child welfare, the elderly, and day care, were categorized under the "welfare" agency. Although past studies do distinguish between "private" or "public" service producers, as Donahue (1989) notes "there is an almost infinite variety of possible organizational forms, with varying mixes of 'publicness' and 'privateness' " (p. 9). Admittedly, the designations used in this study—for-profit, nonprofit, and government—are also subject to the same criticism, but nevertheless represent a step toward associating the retrenchment effects contracting to specific sectors.

Hypothesis I: Property rights hypothesis. Private for-profit organizations produce services at lower costs in comparison with either government or nonprofit organizations. Nonprofit organizations and governments do not enable divisibility of ownership rights into the organization, which motivates inefficiency.

Hypothesis II: Public choice hypothesis. Private for-profit organizations produce services at lower costs than either the nonprofit or governmental sector because of scale economies and greater competition in the private sector. However, nonprofits are comparably more efficient than the governmental sector, because contracting with the latter

TABLE 7.1 Variables and Data Sources

Variables	*Data Source*
Expenditure data	City Government Finances 1981-1982
Employment data	Census of Government Employment 1982
Proportion of services contracted out based on 64 functions	ICMA Survey of Alternative Service Delivery Approaches 1982
Proportion of services contracted out by agency	ICMA Survey of Alternative Service Deliver Approaches 1982
Percentage of 64 goods and services for which the municipality assumes responsibility	ICMA Survey of Alternative Service Delivery Approaches (1982)
City population, 1980	U.S. Bureau of the Census (1982)
Mean per capita income, 1983	U.S. Bureau of the Census (1983)
A dummy variable that equals 1 if the city is located in an SMSA, otherwise, 0	U.S. Bureau of the Census (1982)
A dummy variable that equals 1 if the city uses a council-manager form of government, and 0 if the city has a mayor-council form	ICMA Survey of Alternative Service Delivery Approaches 1982
Percentage of total municipal revenues from intergovernmental grants, 1982	U.S. Bureau of the Census (1982)
Percentage of the municipal work force that is unionized	U.S. Bureau of the Census (1982)
Metropolitan area average monthly wage for a nonpublic worker, 1982	U.S. Bureau of the Census (1982)
Average monthly wage per full-time equivalent municipal worker 1982	U.S. Bureau of the Census (1982)

NOTE: Data from the International City Management Association were merged with Fiscal Austerity and Urban Innovation Project data files that included several U.S. Census sources, mainly the 1980 Census of Population.

amounts to transferring production from one "monopoly government" to another.

Hypothesis III: Hybrid contract failure/subsidy hypothesis. Nonprofits perform services at lower costs than either governmental or for-profit enterprises. Nonprofits are less motivated to take advantage of informational asymmetries facing consumers. Moreover, subsidies provided by other governments and private sources allow them to operate as if there is a reduction in either their fixed or variable costs, enabling them to charge municipalities less.

III. Findings

Does municipal contracting promote retrenchment? Table 7.2 shows that the greater the level of service contracting (i.e., percentage of service contracted out irrespective of sector) the *lower* the level of expenditures and employment, consistent with findings by Ferris (1988). By contrast, the greater the level of contracting the *higher* the average level of wages. This latter finding suggests that either (a) gains from contracting are substituted for higher salaries citywide, or (b) contracting reduces the level of more poorly paid positions, which "automatically" increases the average wage because the employees that are left are more highly skilled (Stein 1990). Controlling for the level of contracting and other factors shown to influence policy outputs in past studies, higher rates of unionization lead to higher expenditures, employment, and average wages. Having shown that contracting does reduce spending, the discussion below returns to the main focus of this study—whether the cost savings effects of contracting are associated with some sectors more than others.

Expenditures

Table 7.3 shows the effects by different sectors of contracting on agency expenditures controlling for the independent variables in Table 7.2. Compared to Table 7.2, the key substitution in Table 7.3 is that the contracting measure refers to percentage of that agency's contracts with a specific sector. The findings are striking. In contrast to the expectations of either the property rights or public choice approaches, contracting with nonprofits promises the greatest potential for retrenchment—other factors constant—followed by contracting with the private for-profit sector. Contracting with government fares the worst.

More specifically, contracting with for-profit firms leads to lower expenditures for highways and health. Contracting with nonprofits leads to retrenchment overall, but also to lower expenditures in 9 of 12 agencies. Only in the instances of libraries and cultural programs did contracting with nonprofits lead to higher expenditures. Contracting with governmental organizations did not generally reduce expenditures. For sanitation, highways, hospitals, and utilities, contracting with governments generally increased expenditures. Only in the case of parks and recreation, where government cooperation

TABLE 7.2 Effects of Contracting Out on Aggregate Expenditures, Employment, and Wages

	Expenditures			*Employment*			*Average FTE Wages*		
Independent Variables	*B*	*SE*	*t*	*B*	*SE*	*t*	*B*	*SE*	*t*
Mean per Capita Income	.206	.056	3.71***	–.203	.131	–1.54	.169	.017	9.78***
City Located in SMSA (1 = Y)	–.074	.036	–2.05**	.563	.086	6.56***	.078	.012	6.62***
Intergovernmental Revenues	.756	.137	5.52***	.150	.325	0.46	–.014	.044	–0.32
Council-Manager Government (1 = Y)	.033	.030	1.10	.276	.070	3.92***	.061	.009	6.44***
Average Monthly Pay (see note below)	.375	.084	4.44***	.320	.200	1.60	.100	.016	6.14***
Percent of Public Employees Unionized	.144	.070	2.04**	.794	.167	4.76***	.380	.020	18.96***
Functional Responsibility Score	1.065	.073	14.51***	2.234	.174	12.85***	–.031	.024	–1.29
Percent Services Contracted Out (64 Functions)	–.050	.013	–3.89***	–.096	.030	–3.15***	.012	.004	3.01**
Constant	–4.109	.755	–5.44***	–5.339	1.789	–2.99***	4.956	.213	23.31***
Number of Cases		1330.00			1330.00			1330.00	
R Square		.20			.20			.43	

NOTE: PC = per capita. Average private sector wage is used to explain public wage levels; public sector wages are used in expenditure and employment regressions. All dependent variables are per capita. * = $p<.10$; ** = $p<.05$; ***=$p<.01$.

TABLE 7.3 Effects of Contracting by Sector on Functional Expenditures

	Sector of Contract			
Function	*For-Profit*	*Nonprofit*	*Inter-governmental*	R^2
Total Expenditures	–.061	–.052**	.066	.52
	(.040)	(.022)	(.041)	
Police Protection	.031	–.048	–.127	.19
	(.224)	(.042)	(.238)	
Fire Prevention	.122	–.719***	–.334	.30
	(.404)	(.069)	(.415)	
Sanitation	–.087	–.183**	.309*	.09
	(.126)	(.080)	(.163)	
Sewerage	.099	–.301**	–.198	.19
	(.634)	(.131)	(.648)	
Highways	–.210**	–.146***	.224**	.14
	(.086)	(.049)	(.109)	
Transit	–.468	–1.063***	.090	.22
	(.302)	(.145)	(.345)	
Parks & Recreation	1.305***	–.242***	–1.530***	.31
	(.491)	(.054)	(.506)	
Libraries & Cultural Programs	.051	.428***	—	.15
	(1.046)	(.071)		
Health	–.279*	–.159***	.147	.25
	(.169)	(.052)	(.172)	
Hospitals	–1.508	–.939***	2.766***	.18
	(.986)	(.220)	(1.023)	
Housing	–.332	–.365**	.393	.14
	(.619)	(.186)	(.649)	
Welfare	.216	–.088	–.293	.34
	(.461)	(.069)	(.463)	

NOTE: Standard errors are in parentheses. $N = 1333$. Control variables are shown on Table 7.1.
* = $p < .10$; ** = $p < .05$; *** = $p < .01$.

(e.g., city/county agreements) can benefit from scale economies, did contracting reduce expenditures. Besides Hypotheses I and II, another reason for the relatively poorer performance with governmental contracts may be the *motivation* for the contract itself. In some instances, municipal managers may care more about making sure the service is provided (e.g., in the instance of a state mandate) than the actual costs of providing it, especially if they do not have the specialization to provide it in-house (DeHoog 1986).

TABLE 7.4 Effects of Contracting by Sector on Functional Employment

	Sector of Contract			
Function	*For-Profit*	*Nonprofit*	*Inter-governmental*	R^2
Total Employment	−.017	−.208***	−.004	.29
	(.116)	(.064)	(.119)	
Police Protection	.532	.231***	−1.637***	.28
	(.444)	(.083)	(.471)	
Fire Prevention	.491	−1.434***	−.433	.36
	(.592)	(.101)	(.608)	
Sanitation	−.644***	−.295***	.945***	.22
	(.180)	(.114)	(.233)	
Sewerage	−.024	−1.093***	.026	.33
	(.698)	(.144)	(.714)	
Highways	−.127	−.298***	.182	.22
	(.140)	(.080)	(.177)	
Transit	−1.104***	−.961***	.741***	.24
	(.237)	(.113)	(.269)	
Parks & Recreation	2.404***	−.369***	−2.602***	.32
	(.684)	(.076)	(.706)	
Libraries & Cultural Programs	1.115	.433***	—	.18
	(1.267)	(.087)		
Health	−.224	−.186***	−.089	.24
	(.199)	(.061)	(.203)	
Hospitals	1.255	−.113	−1.396	.15
	(1.013)	(.226)	(1.051)	
Housing	.503	−.064	−.413	.20
	(.467)	(.140)	(.490)	
Welfare	−.643	−.138	.456	.19
	(.592)	(.088)	(.595)	

NOTE: Standard errors are in parentheses. $N = 1333$. Control variables are shown on Table 7.1.
* = $p < .10$; ** = $p < .05$; *** = $p < .01$.

In summary, the aggregate expenditure reduction effects of contracting shown on Table 7.2 can be attributed mostly to contracts with the nonprofit and, to a lesser extent, for-profit sectors.

Employment

Table 7.4 shows the equations regressing contracting and other independent variables on aggregate and functional employment levels. Once again, contracting with nonprofits has a statistically significant negative impact on aggregate employment. Although the coefficients for contracting with for-profits and governments are in the negative direction, they are statistically insignificant. For 2 of 12 functions, contracting with for-profits has a statistically significant negative effect on employment levels. In the instance of intergovernmental agreements, contracting has a statistically significant negative effect on only 2 of 12 agencies and has a significant positive effect (i.e., costs more than maintaining the service in-house) in as many agencies. Once again the striking pattern appears to be with nonprofits—statistically significant negative effects of contracting on employment are found in 7 of 12 agencies. Only in the instances of libraries, cultural programs, and police did contracting with nonprofits actually increase employment levels.

IV. Conclusions

What does the empirical evidence suggest with respect to the three hypotheses stated above? First, as Bozeman (1987) and Donahue (1989) suggest, it is important to recognize that no study can make a universal assertion about the relative efficiencies about particular sectors. Too many intervening variables (e.g., scale economies, market structure, transaction costs) can confound validity. With that qualification in mind, our findings do not support Hypothesis I—that divisibility of property rights is the principal determinant of efficiency. Contrary to the expectations of property rights theory, nonprofits, followed by for-profits, and then government production is the ranking to be followed by municipal officials seeking to gain cost efficiencies from contracting. The main assumptions in this assertion are that quality differences between sectors are negligible and the nonprofit sector has the technological capacity to produce specific services.

What about the public choice approach (Hypothesis II)? The public choice approach is also generally unsupported by our findings to the extent that the "privateness" of an organization necessarily matters. The public choice tradition predicts that for-profit contracting promises the greatest cost savings because of competition and the residual claimant status associated with this organizational form. Yet our findings are potentially consistent with many public choice studies on the importance of competition. Nonprofits may be producing services at lower costs because of the more intense competition among them in particular functional areas (e.g., housing and hospitals).

Our findings are most supportive of Hypothesis III—the contract failure/subsidy hypothesis. However, sorting out the specific reasons why nonprofits achieve comparably greater cost efficiencies in several functional areas is more difficult to assess with our data, and thus merits further empirical inquiry. Nevertheless, several possible reasons for greater cost savings associated with nonprofit contracting are discussed below.

Nonprofits may indeed be less likely to take advantage of *informational asymmetries* and raise contract prices during "ex-post" contracting than for-profits. However, further empirical evidence is needed to corroborate this point. Nonprofits have a *more diverse revenue mix.* When they bid to produce a service where municipalities have a role, they may receive "piggyback" funding from states, the federal government, and private sector foundations that allows them to charge municipalities less (Bacon 1989).

Nonprofits can also charge municipalities less because of the *volunteer labor* that is used in their input mix. As Peter Drucker (1989) notes:

> Few people are aware that the nonprofit sector is by far America's largest employer. Every other adult—a total of 80 million plus people—works as a volunteer, giving on average nearly five hours each week to one or several nonprofit organizations. . . . Were volunteers paid, their wages, even at minimum rate, would amount to some $150 billion, or 5% of GNP. (p. 88)

The cost savings from nonprofits may also result from the *more competitive market structure* of nonprofits. It may well be more competitive than that for many for-profit or governmental services. Situations in which municipalities keep some of the service in-house and contract out the rest to nonprofits may introduce "benchmarks" that improve the efficiency of both nonprofits and government agencies (Miranda 1993).

Finally, another reason—put forth by Ferris and Graddy (1991)—is *transaction costs.* More specifically, municipalities contracting with nonprofits may incur fewer expenditures in monitoring nonprofits on service quality specifications. Thus, it may not be the more efficient performance of nonprofits per se, but the possibility that municipalities expend fewer resources on contract monitoring and implementation.

In summary, the results from this study *do not* demonstrate that the organizational form of nonprofits is more efficient than others. Contracting with nonprofits may be associated with cost savings because of subsidization either directly (e.g., grants and cash contributions) or through in-kind services (e.g., volunteer labor). However, the implications for municipal managers are relatively straightforward. Contracting with nonprofits can promote cost savings. And the savings are larger than contracting with for-profits or other governments. By contrast, the conventional wisdom is that nonprofits are more attractive partners in service delivery because they take greater *care* and provide higher *quality*.

References

Ahlbrandt, R. 1973. "Efficiency in the Provision of Fire Services." *Public Choice* 16:1-15.

Alchian, A. 1965. "The Basis of Some Recent Advances in the Theory of Management of the Firm." *Journal of Industrial Economics* 14:30-41.

———, A. and Demsetz, H. 1972. "Production, Information Costs, and Economic Organization." *American Economic Review* 62:777-95.

Bacon, D. 1989. "Nonprofit Groups: An Unfair Edge?" *Nation's Business* 77(4):33-34.

Bays, C. 1979. "Cost Comparisons of Nonprofit and Nonprofit Hospitals." *Social Science and Medicine* 13C:219-225.

Becker, E. and F. Sloan. 1985. "Hospital Ownership and Performance." *Economic Inquiry* 23(1):21-36.

Bennett, J. and M. Johnson. 1979. "Public Versus Private Provision of Collective Goods and Services: Garbage Collection Revisited." *Public Choice* 34:61-62.

Berenyi, E. and B. Stevens. 1988. "Does Privatization Work? A Study of the Delivery of Eight Local Services." *State and Local Government Review* 20(1):11-21.

Borcherding, T., ed. 1977. *Budgets and Bureaucrats: The Sources of Government Growth.* Durham, NC: Duke University Press.

———, B. Burnaby, W. Pommerehne, and F. Schneider. 1982. "Comparing the Efficiency of Private and Public Production: The Evidence From Five Countries." *Journal of Economics* 2:127-56.

Bozeman, B. 1987. *All Organizations Are Public.* San Francisco: Jossey-Bass.

Coase, R. 1937. "The Nature of the Firm." *Economica* 4:386-405.

———. 1960. "The Problem of Social Cost." *Journal of Law and Economics* 3:1-44.

Deacon, R. 1979. "The Expenditure Effects of Alternative Public Sector Supply Institutions." *Public Choice* 33:381-98.

De Alessi, L. 1969. "Implications of Property Rights for Government Investment Choices." *American Economic Review* 58:13-24.

DeHoog, R. 1986. "Evaluating Human Services Contracting: Manager, Professionals, and Politicos." *State and Local Government Review* 18:37-44.

Donahue, J. 1989. *The Privatization Decision.* New York: Basic Books.

Drucker, P. 1989. "What Business Firms Can Learn From Nonprofits." *Harvard Business Review*. July-August (4): 88.
Ferris, J. 1988. "The Public Spending and Employment Effects of Local Service Contracting." *National Tax Journal* 41(2):207-17.
——— and E. Graddy. 1986. "Contracting Out: For What? With Whom?" *Public Administration Review* 46:332-44.
———. 1988. "The Production Choices for Local Government Services." *Journal of Urban Affairs* 10:273-89.
———. 1991. "Production Costs, Transaction Costs, and Local Government Contractor Choice." *Economic Inquiry* 29:541-54.
Hansmann, H. 1987. "Economic Theories of Nonprofit Organization." Pp. 27-54 in *The Nonprofit Sector: A Research Handbook*, edited by Walter W. Powell. New Haven, CT: Yale University Press.
Kemper, P. and J. Quigley. 1976. *The Economics of Refuse Collection*. Cambridge, MA: Ballinger.
Migue, J. and G. Belanger. 1974. "Towards a General Theory of Managerial Discretion." *Public Choice* 17:24-43.
Miranda, R. A. 1992a. "Privatization and the Budget Maximizing Bureaucrat." Presentation at the Association of Public Policy Analysis and Management (APPAM) Conference, September, Denver, CO.
———. 1992b. "Privatization in Chicago's City Government." *Research in Urban Policy*, 4:31-53. Greenwich, CT: JAI.
———. 1992c. "Privatizing City Government: Explaining the Adoption and Budgetary Consequences of Alternative Service Delivery Arrangements." Ph.D. dissertation, Harris Graduate School of Public Policy Studies, University of Chicago.
———. 1993. "Bureaucracy, Organizational Redundancy and the Privatization of Public Services." Presented at the Berkeley Public Management Conference, July.
———. 1994. "Governments or Markets? The Privatization of Municipal Services." *Research in Governmental and Nonprofit Accounting* 8. Greenwich, CT: JAI Press.
Nelson, R. and M. Krashinsky. 1973. "Two Major Issues of Public Policy: Public Policy and Organization of Supply." In *Public Subsidy for Day Care of Young Children*, edited by Richard Nelson and Dennis Young. Lexington, MA: D.C. Heath.
Niskanen, W. 1971. *Bureaucracy and Representative Government*. Chicago: Aldine Atherton.
O'Neill, M. 1989. *The Third America*. San Francisco: Jossey-Bass.
Osborne, D. and T. Gaebler. 1992. *Reinventing Government*. Reading, MA: Addison-Wesley.
Ostrom, V., C. Tiebout, and R. Warren. 1961. "The Organization of Government in Metropolitan Areas: A Theoretical Inquiry." *American Journal of Political Science* 55(4):831-42.
Robinson, J. and H. Luft. 1985. "The Impact of Hospital Market Structure on Patient Volume, Average Length of Stay, and the Cost of Care." *Journal of Health Economics* 6:333-56.
Salamon, L. 1992. *America's Nonprofit Sector: A Primer*. Baltimore, MD: Johns Hopkins University Press.
Savas, E. S. 1987. *Privatization: The Key to Better Government*. Chatham, NJ: Chatham House.
Stein, R. 1990. *Urban Alternatives*. Pittsburgh, PA: University of Pittsburgh Press.

Watt, J., S. Renn, J. Hahn, R. Derzon, and C. Schramm. 1986. "The Effects of Ownership and Multihospital System Membership on Hospital Functional Strategies and Economic Performance." Pp. 260-84 in *For-Profit Enterprise in Health Care*. Washington, DC: National Academy Press.

Weisbrod, B. 1988. *The Nonprofit Economy*. Cambridge, MA: Harvard University Press.

Wilson, G. and J. Jadlow. 1982. "Competition, Profit Incentives, and Technical Efficiency in the Provision of Nuclear Medicine Services." *Bell Journal of Economics* 13(2):472-82.

8

Innovations That Work

A Menu of Strategies

Terry Nichols Clark

Introduction

How do cities respond to fiscal constraints and austerity? The short answer is, with huge diversity. The specifics, however, are seldom assembled and compared systematically. Most city officials do not communicate their policies beyond their borders; they implement them locally. Yet what seems mere "common sense" in one locale can be wisdom for others.

The keen interest of mayors to manage cities better emerged in workshops Terry Clark helped arrange during a period of several years, often sponsored by the U.S. Conference of Mayors. This led the conference to invite him to prepare a book about city leadership and fiscal management innovations. John Gunther, executive director of the conference, invited mayors to submit interesting ideas and documents. Nearly 100 did. Terry Clark visited and met with many of their officials. The first results were published as brief discussions of innovative practices in U.S. cities (in T. N. Clark, DeSeve, and Johnson 1985). Some local officials asked, "Can I use these ideas in my city?" Analysts pose a more general version of this same question: How to explain where and why policies are adopted?

We developed long lists of new ideas that cities were trying, and classified them in various ways. How to assess innovativeness of an individual city? Abstract writing on this question is inconclusive.[1] Many writers seek to pigeonhole policies, but leave unanswered how to characterize a city in its overall profile of policies. Scanning related work by others, we felt some of the most challenging was by Richard Bingham and Brett Hawkins. Their insight was to offer city officials a list of policies, ask which they were using, and follow up with more specific questions. The three of us developed a list of about 30 strategies which some cities were adopting. We tried these in pretests with city officials, sent the questionnaire to many specialists for reactions, and eventually chose "the infamous 33" strategies for the survey of the Fiscal Austerity and Urban Innovation (FAUI) Project. The 33 appear in Table 8.1, with results for the 1983-1984 main FAUI survey and three updates for the United States.

How to interpret the strategies? Three main approaches have evolved in the FAUI Project:

1. Study *individual strategies*, such as contracting out, for interesting variations and creative twists.
2. Consider *patterns of strategies* that go together empirically, via statistical procedures like factor analysis.
3. *Group strategies* in ways that correspond *to a more general analytical perspective.*

We use all three approaches below (and review past related work).

City Strategies to Alleviate Austerity and Promote Innovation

By strategy we refer to distinct sets of policies that many cities adopt, some of which are innovative and others not at all. Which strategies deserve the label of innovation? In our view definitions are arbitrary, and what counts most is what is both original and useful to an individual city for a given problem at a given time. This approach to innovation is relativistic and local, but it still can incorporate more general criteria. Here for instance are five criteria that were applied by the judges in choosing among applications in mak-

TABLE 8.1 Local Choices of Fiscal Austerity Strategies

Fiscal Austerity Strategy	FAUI 1983-84 Percent Cities Using	Walzer/Deller 1987 Rank in Use by Counties 1= most frequent	NLC 1992 Percent Cities Using	Brooks 1992 Rank in Use by Ohio Cities 1=most frequent
Increase user fees	85	5, 9*	54	4
Improve productivity by management techniques	75		34*	1
New local revenue sources	73	2,3,4*		3
Attrition reduction of workforce	71			7
Reduce spending on supplies, etc.	68			11
Draw down budget surplus	64			13
Improved productivity by labor saving techniques	63		34*	2
Increase taxes	61	2,3,4*	24	14
Hiring freeze	59	11	44	15
Reduce capital expenditures	58			9
Across-the-boards budget cuts	56			10
New intergovernmental revenue	54			5
Reduce Overtime	53			8
Reduce administrative expenses	52	12		
Keep expenditure increases below inflation	51			6
Eliminate programs	48			
Contract services out to private business	46		31*	17
Joint purchasing agreements	46		23*	11
Layoffs	46	14	40	21
Sell assets	41			22
Reduce locally funded services	40		14	
Cut least efficient departments	39			16
Defer maintenance	37	7		
Reduce intergovernmental funded services	37			
Increase long-term borrowing	36	10		18
Defer payments	35			20
Increase short-term borrowing	34			17
Shift responsibilities to other governments	32		3	
Salary freeze	31	11*		
Contract services out to other governments	29		31	
Early retirements	26			
Reduce pay levels	23			
Controls on new construction to limit population	19			

SOURCES: 1. U.S. Fiscal Austerity and Urban Innovation Project, $N = 517$. Item "Here is a list of fiscal management strategies that cities have used. Which have you used since January 1978? . . . Please indicate the importance in dollars of each strategy (since January 1978): One of the most important, Very important, Somewhat important, Least important, Don't know/not applicable." Subsequent studies: 2. Walzer and Deller (1993) surveyed 553 U.S. rural county highway officials about strategies likely to be pursued in response to termination of General Revenue Sharing over the next two years. Key differences: taxes include three items: property, sales, and motor vehicle or wheel tax; user fees include service fees and charges and private development fees. 3. Pagano (1993) reports a 1992 survey by the National League of Cities of 620 cities. Key differences: NLC asked for strategies used in last 12 months (thus lowering all percentages compared to FAUI items). Productivity was one item only: "Improved productivity levels," and only "contract out services" was asked instead of the two FAUI items. 4. Brooks (1993) was a replication of the full FAUI survey in 22 Ohio cities; all 33 strategies asked; infrequent responses omitted here.
NOTE: * = item not directly comparable.

ing more than 70 awards for the Urban Innovation in Illinois program (see Urban Innovation in Illinois 1988, 1989):

1. Potential for applicability and continued use in other governments
2. Innovativeness and originality (a) nationally and (b) in Illinois
3. Improved service quality—better service
4. Degree of productivity improvement—increased service per dollar
5. Creative implementation—such as winning support among potential opponents to the idea

We feature below several examples of innovations that meet these criteria. Consider the 33 strategies first in the following groups: revenues and debt, cutting surpluses, borrowing, cuts, productivity, joint agreements among governments, and contracting out.

Revenues and Debt

The major revenue changes of U.S. cities during the past two decades have been:

- Cuts in federal and state grants
- Less use of taxes, especially property taxes
- More user fees and miscellaneous revenues
- New forms of debt

We consider each below, except for grants, as these are detailed in Chapter 6, and analyzed further in Chapters 2 and 5.

Taxes Have Declined, Especially Property Taxes

Views of government shifted dramatically in the 1970s, when taxpayer revolts ended decades of growth. If taxpayers do not all appreciate how much taxes have fallen, city officials do. The situation is especially dramatic in contrasting the 1990s and 1970s. Example: property taxes not collected in California for more than a decade solely because of the passage of Proposition 13 in 1978 amounted to $228 billion—more than the annual federal deficit (calculation by California Taxpayers' Association cited in D. Jeffe and S. B. Jeffe 1988, p. 19). Citizens in some states, such as Massachusetts and Michigan, also imposed new tax limits via referenda. But in other places, such as Illinois, tax cuts were made by elected officials following public resistance to taxes. Such "informal" policies set by elected leaders, in aggregate, are far more critical than formal laws and tax limits, even if the informal policies get less media coverage. Officials across the United States have drastically changed their fiscal priorities since the 1970s. A turning point was 1974, when half of the cities began trimming their local expenditures (T. N. Clark and Ferguson 1983). These cuts went unrecognized by journalists and most analysts at the time. But "cutback management" became dramatically visible when California citizens passed Proposition 13 in 1978. Thereafter, if a city council discussed a major spending or especially property tax increase, it was commonly assessed as to whether it might generate a "Prop. 13-type" reaction among taxpayers (T. N. Clark 1990). The consequence: all taxes declined by about a quarter from 1970 to 1990, and property taxes fell from 48 to 31 percent of cities' own source revenues (these and similar results below come from our calculations using Census fiscal data).

Property taxes have been criticized for decades by many economists and policy advisers as economically regressive and thus undesirable for poorer residents. But paradoxically, in the very years that citizens so turned against them, Chicago economist Arnold Harberger put forth a new analysis: the traditional interpretation of property tax incidence was wrong, he argued; property taxes are not regressive, but progressive. Among some policy advisers, this in turn weakened past arguments for property taxes.[2]

User Fees Are More Popular

User fees often fill gaps left by cuts in intergovernmental grants and property taxes. In many states, user fees are exempt from legal limits like Prop. 13 on property and other taxes. Fees thus jumped dramatically in California the year after Prop. 13, and kept increasing.[3] California was visible but not alone. Property taxes were roundly attacked across the United States in the mid-1970s, and local officials seeking to maintain services often shifted to user fees plus a range of other "less visible" revenues. User fees ranked as the most frequent strategy chosen by U.S. cities in 1983-84 as well as in the National League of Cities (NLC) 1992 update. They were slightly lower in the two other surveys in Table 8.1, which were of smaller localities. Fees illuminate U.S. values. They are widely deemed "fairer" than taxes because each user pays for services received—marginal supply is thus matched with marginal demand, and "economic efficiency" better realized, the clinching argument in neoclassical economics. Fees also provide information about citizen preferences for specific services (Mushkin and Vehorn 1980). These aspects of fees appeal to U.S. individualism, which in the 1980s swamped the egalitarian argument that all citizens should receive the same service. Some evidence (Rigos 1986) suggests that fees are adopted more often by fiscally healthy cities with a significant middle-class citizenry that prefers fees to taxes. Fees and charges rose from 27 to 39 percent of own source revenues for all U.S. cities in the 1970 to 1990 period. Compare their growth to that of property taxes: in 1970 fees generated only 56 percent as much revenue as property taxes, but this more than doubled by 1990, when they generated 126 percent as much revenue as property taxes.

Some municipalities rely heavily on charges. For example, in Hawthorne, Los Angeles County, California, charges generate more than twice as many revenues as the property tax. The Chicago suburb of Schaumburg boasts zero property taxes, but raises substantial revenues from water and electricity fees (T. N. Clark et al. 1989). In small, homogeneous locales with few minorities, the individualistic arguments may count for more than in big cities—traditionally Democratic and heavily minority and poor.

These are broader ideological arguments into which user fees fit. But the fit is not always firm. Important exceptions to U.S. individualism include Democratic strongholds like Chicago, where egalitarian arguments usually win in council floor debates. Yet in actual fiscal policies, Chicago uses fees for the majority of its revenues—charges on electricity and garbage generate more revenue than property taxes.

Other reasons for fees are bureaucratic. Fees can help the department providing the service: if citizens pay for garbage pickup, the sanitation department often receives the "benefits." Many states restrict fees to the cost of service (MacManus 1983, p. 160). Yet because "actual cost" involves some discretion, department heads may use fees to display their revenue-raising talents, expand their organizational domains, and advance their careers. This is the expansive bureaucrat argument about the growth of government (see Chapter 5). On the other hand, if the central budget office prohibits departments from holding on to any of their new revenues or cost savings, this discourages innovation by department heads and lower level staff.[4] Separate departments, such as the budget office or productivity agencies, must thus work harder and may find their suggestions unwelcome if they are resented by basic city staff.

New Revenue Sources

"New revenue sources" ranked just below user fees in our FAUI survey. These revenues are a mixed lot, sometimes deliberately hidden under "miscellaneous" in financial reports by officials seeking untapped revenues that will not rouse popular resistance. California municipalities worked particularly hard to tap unusual sources just after Prop. 13 in 1978 and again following a general spending limit the next year. These placed a premium on finding revenues outside the legal limits. Example: "economic development" impact fee packages mushroomed, with developers assessed special (expensive!) fees to acquire licenses. In some instances developers even provide a revenue stream from profits on enterprises like shopping centers to the local government as a shareholder. Chicago's Mayor Jane Byrne imposed dozens of new fees and raised old ones, such as for the sticker required for all motor vehicles, thereby generating a small public furor. Still, many leaders and their advisers would chose fees—indeed almost any other revenue source—over property taxes.

Cutting Surpluses and Selling Assets: "Leaner Management" or "Irresponsibility"?

Cities cannot print money or carry deficits like the federal government, but the strategies of drawing down surpluses and selling assets are controversial and often considered irresponsible. They were still reported as used by 64 and 41 percent respectively of U.S. cities in our FAUI CAO survey.

Reducing Surpluses. Reducing surpluses is really not inherently "bad"; professional opinion is mixed. Keeping a large surplus in a "rainy day fund" may look like "sound management," but revenues can be saved only by collecting more than is needed to spend. Carrying a smaller surplus demands more precise monitoring of revenue and expenditure flows to avoid a deficit. But this is like managing a private firm with a smaller inventory—heralded as efficient management in the 1980s, when it became a real option due to more sophisticated computer procedures for closer monitoring.

Selling Assets. Selling assets can indicate simply cleaning out warehouse "junk." But sale of land or buildings, such as the controversial land sales in downtown Chicago and New York City under fiscal pressure in the 1970s, represents a one time only "depleting of municipal assets." Is this salutary slimming to achieve "leaner government"? When San Diego sold attractive parkland to raise revenues, the sale roused dissent among ecology-conscious Californians who considered it a scandalous sacrifice of the "commonweal."

We did not inquire about more complex practices like transferring monies across separate funds to avoid reporting deficits, but this and many other practices have flourished as finance staff are pressed by mayors and council members to "find the money, anywhere"—without "imposing new revenues." Lennox Moak was a famous case in his last years as finance director of Philadelphia in the 1970s.[5]

These examples illustrate how difficult "professional judgment" is to define, and that one should especially not conclude that city staff are irresponsible—professional staff often caution against more questionable practices.

Even the Lennox Moak scandal was tame compared with two dramatic examples from Europe, from two countries supposedly dominated by centralized national bureaucracies. One local official from an innovative French city reported how he increased returns using the city's cash. French cities, like those in the United States, are officially prohibited from many investments, such as common stocks. To circumvent this rule, this city advanced cash to a sports club subsidized by the city; the club in turn invested in the Paris stock market and generated attractive returns for the city. This stratagem was proudly recounted to other local officials in a public meeting attended by Ministry of Interior staff (and some FAUI participants)! Second, in Sweden, a young local official invested millions in options and lost much of the capital. He was fired, and the case became a national scandal.

As Durkheim observed, basic norms are revealed most clearly when they are violated and punishments are meted out. Although the United States is more decentralized than Europe in some respects, these examples suggest that in some cases U.S. professional associations like the Government Finance Officers Association, and behind them the courts and legal system, provide more national constraints on "innovations" in the United States than in more "state-regulated" European countries.

Borrowing: New Modes Cost Less

With cuts in grants, and resistant taxpayers, debt has become a more important option for cities to review carefully. For decades cities borrowed in "plain vanilla" manner using long-term (10- to 30-year), fixed-rate municipal bonds. But the massive stagflation of the 1970s saw interest rates go so high that many cities could no longer borrow at all. Total city debt outstanding also fell 30 percent from 1970 to 1975, the very years in which grants and taxes dropped.

Some innovative governments—such as the State of Michigan and cities of New York, Boston, and Chicago—then fashioned new ways to borrow. For example, very short term borrowing (repaid daily or weekly) was possible well below interest rates for long-term debt. Some cities thus borrow large amounts over short terms that continue for years. "Enhancements" such as pledging a specific revenue or having a commercial bank or insurance company guarantee the loan can further reduce interest costs. "Derivatives" are a more recent development that can involve combinations of transactions such as swapping long-term fixed-rate payments for short-term variable payments. Derivatives allow city governments to create synthetically a wide variety of bond structures and to achieve advantages—such as lower rates or risk or a longer term—not possible with a single instrument. Such developments in the municipal debt market mirror the complexities of corporate finance in these years. Many such strategies are developed continuously by investment bankers. Although complex, these new approaches can cut borrowing costs by half. Many other devices entered debt management in the 1970s and 1980s, making it a far more sophisticated area than earlier, as illustrated by the change since 1972, when 75 percent of long-term issues were priced by taking competitive bids. By 1992, the complexity had led to only 25 percent being sold by competitive bids (Ferguson 1993). Prices are instead negotiated with a limited group of bankers who provide more specific services. Debt outstanding from 1975 to 1990 has increased at about the same rate as city-owned source revenues.

Innovations come in the new forms of borrowing and proceed use—such as lending money borrowed at public sector rates to a private firm, which constructs buildings, which are leased back to the city. Or making loans to private developers at below-market rates for hotels or shopping centers in return for a share of revenues. The most extreme forms of such activity used the Industrial Development Bonds that were largely outlawed in the 1986 Tax Reform Act.

Professional financial advisers and investment bankers have played important roles in shaping these new tools for municipalities. They actively visit local finance officials in search of business, summarizing alternatives and competing with each other, making it easier for city staff to stay abreast of such developments than in many other municipal policy areas. But many financial innovations are too risky or expensive for many municipalities. How localities can develop their staff's professional talents and use the right combination of outside advisers remain difficult questions; city officials can often exchange notes on these issues to mutual benefit.

Cuts: Reduce Expenditures Through Personnel Actions

Cuts in the 1980s were especially dramatic in areas such as Alaska, where oil generates some 90 percent of state revenues and local governments receive half their revenues from the state. In the mid-1980s, oil prices dropped from more than $30 per barrel to about $12, generating immediate and drastic cuts for local governments. Alaska's second largest city, Fairbanks, faced the hard choice of finding more local revenues or cutting staff. The council was besieged by citizens whose jobs and business had also been cut; they vociferously opposed any taxes or revenue increases. Highly unionized and militant, city employees lobbied hard, but lost: the council approved no new revenues and staff was cut about 40 percent in less than one year.

Similar cuts were imposed in other oil-dependent areas, such as Nigeria: more than 50 percent of staff and almost all capital projects were cut in Lagos. But in Norway, a large "rainy day fund" from oil revenues permitted increases until 1986, when cuts finally came (Baldersheim 1992, pp. 83ff).

City officials are seldom proud of the eight strategies we labeled "Reduce Expenditures Through Personnel Actions," such as layoffs, hiring freezes, and so on (see Table 8.1). An exception is when they may be part of a package, that is, "despite staff cuts of 30 percent, we maintained the same park services." Yet this is less a cut than a productivity improvement, as discussed below.

Cuts: Reduce Expenditures "Without" Personnel Actions

Under the heading of reducing expenditures without personnel actions fall 9 of our 33 strategies that cut expenditures but (not directly) personnel: cuts in all departments, decrease capital expenditures, and so forth (see Table 8.1). These were common in most U.S. cities. Yet most city officials consider cuts "uninteresting" to report to other cities. More appealing is cutting expenditures, but not services. These are "productivity" strategies, which seek to "do more for less."

Productivity: More for Less

In doing more for less emerges the genius of local autonomy: with thousands of local governments confronting fiscal constraints, many respond creatively. Some may fail, or simply change little, but many succeed in genuinely improving productivity, defined simply as doing more with less, providing more or the same service level for less cost. The international Fiscal Austerity and Urban Innovation Project supports many national and regional subprojects. One is Urban Innovation in Illinois (UII), which awards prizes to local officials for innovative policies they develop. UII then conducts case studies documenting the innovations and helps spread information to other local officials via professional meeting presentations, short memos and papers, and informal meetings. A few examples:

Joint Contracting. The village of Lombard, Illinois, noted that staff from the private utility, Nigas, might save them money by reading both Nigas's gas meters and the village's water meters simultaneously. Nigas agreed. Village staff stopped reading meters. The dollar saving was immediate and dramatic and could be generalized to other locales.

The arrangement was so successful that it spread to some 40 other municipalities in Northern Illinois (with help from Urban Innovation in Illinois). This meter case is so "obvious" that one asks why not earlier? The most common barriers we have found to innovation are political and administrative feasibility: if there is no direct encouragement, and if there are some direct *dis*couragements, innovation lags. The lion's share of innovation is not the "technology," but the creative process of matching available technologies with local institutional arrangements. Example: the City of Chicago is classically cited for inefficiency. How to innovate in such a hostile context?

Shared Savings in Chicago. Charles Williams, head of energy for Chicago, saw dramatic conservation potentials in city buildings. But when he sought funding for insulation, more efficient lighting, and so forth, it was denied by the council. How to overcome inertia, skepticism, and a cash shortage? "Shared savings" are a valuable tool for many cities. Williams issued a request for proposals asking firms to analyze energy saving potentials in city buildings, purchase the necessary equipment, and install it at their expense. The firm would be paid when, and only when, city energy bills fell. Firms could then "share the savings," in a fixed proportion to reimburse them for their up-front capital and labor costs. Perhaps the same work might be done at less cost if the outside firm were not asked to front the costs. But for some fiscally pressed cities like Chicago, this is not a "feasible" option. The alternative is often no action at all. Shared savings then merit review.

Contracting Out: Extragovernmental Agreements

Our international FAUI survey included contracting out services to other governments, contracting out to the private sector, and joint agreements among governments for purchasing or service delivery. These have been popular means to improve productivity. A few highly successful examples:

Joint Government Agreements. Northern Illinois municipalities have been national leaders in joint purchasing and service agreements for some two decades.[6] Smaller towns work together in regional associations to purchase rock salt for winter roads, form mutual self-insurance pools, and even joint personnel recruitment and training for new police officers. Savings are obvious and substantial.

An Ad Hoc Joint Municipal Consortium. Several municipalities near Chicago had problems buying trees. Prices skyrocketed for all trees except seedlings. Private nurseries complained that municipalities provided minimal advance notice, especially for large orders. So prices rose. Seven municipalities then created the Suburban Tree Consortium. They calculated a base number of trees they would need for several years and negotiated large volume, multiyear contracts with local nurseries that then planted trees years in advance, knowing that their sale was secured. Result: substantial cost savings.

Computers Yes, Taxes No. Cook County, Illinois, processes the largest number of transactions involving land and building records of any

BOX 8.1 Innovative Performance Indicators: Citizen Surveys

Downers Grove, Illinois (a Chicago suburb), apparently conducts the most systematic citizen surveys of any city in the world. They help officials decide if citizens care about different services. Surveys are one of many performance indicators used to monitor and critique its programs, to make them more efficient and responsive. Downers Grove officials (an assistant manager and a secretary) have conducted mailed surveys of citizens every year for some 15 years, asking detailed questions, such as how courteous are the librarians or school crossing guards and how frequently do water drains back up on your street. If librarians' courtesy ratings drop from 96 to 82 percent in a year, this is noticed. At budget time and in salary decisions, staff know that these results enter the decision process. Listen to staff from Downers Grove talk about their work; they convey a remarkable commitment, responsiveness, and conscientiousness.

Urban Innovation in Illinois (UII) was created with support from the Joyce Foundation and much volunteered aid from local officials. It has made more than 70 awards to Illinois local governments for innovations that reduce fiscal strain by improving productivity. On Downers Grove see Im (1987), subsequent annual reports from the Downers Grove village manager, and Urban Innovation in Illinois (1989).

Examples of survey items used in Downers Grove:

	1992	*1991*	*1990*	*1989*	*1988*	*1987*
Please rate your level of satisfaction with the courtesy and service you received from the Village Water Billing Department.						
Courtesy	79	73	77	71	77	67
Service	80	73	75	68	79	71

(Note: 397 or 65% of respondents have not called or visited the water billing department during 1992.

Please rate your level of satisfaction with the collections of the Library	83	82	82	82	85	86
Please rate your level of satisfaction with the courtesy you received from the Library staff						
Helpfulness	87	88	87	88	89	88

SOURCE: Kurt Bressner. 1992 Citizen Survey Results. Memorandum to Mayor and Council. Downers Grove, Illinois.

local government in the United States. It had a staff of 308 in the mid-1980s, which entered all data by hand and had fallen more than a year behind. The county board denied repeated requests to purchase computers, so the recorder found a private contractor, Business Records, Inc., that agreed to start for free: to install computers and write custom software. The contractor visited Los Angeles County, second largest in volume and boasting the best system there was at the time. For Cook County, Business Records built an even better system. They also maintained the facility with Business Records staff, to train and provide continuing technical support for county staff. Payment? Nothing from property taxes. New fees (paid by persons requesting title information) completely supported the new facility (see T. N. Clark 1988). Satisfaction by local officials: total.

Build a Park and Make Money. Elgin, Illinois, purchased some 400 acres of land for a major park, but citizens were reluctant to spend tax money to develop it. City officials found a creative means to build a park at zero cost to taxpayers. The secret? Not oil, but gravel. A potential quarry lay below the would-be park. So the staff arranged a barter. The city contracted with a private firm to dig the gravel; the firm in exchange developed the park. The firm built an access road and parking lot for the softball complex (a value of $364,000) in exchange for the first 2 years of mining operations. After 2 years, the city received 10 percent of the gross gravel sales—about $75,000 annually for an estimated 10 years. The park grows as gravel is removed. After an area is mined, the operating company does final grading work and plants vegetation. When an area is replanted, it is turned over to the city for park development (see Malme undated; Urban Innovation in Illinois 1989, p. 7).

A New Mayor Often Redefines What Options Are Feasible. Nîmes, France dramatically illustrates how much policies can change with new political leadership. For 18 years, Nîmes was governed by a Communist mayor and council with strong union support. Nîmes is a popular tourist town, and the streets grew littered each weekend, a situation exacerbated by municipal staff that refused to work after Friday morning. Then a new mayor was elected, CEO of Cacharel, France's largest clothing designer company. He brought a private sector approach and contracted street sweeping for the downtown to a private firm. Striking workers, relegated to cleaning streets only in outlying areas, occupied his offices. But the private firm kept the streets clean through the entire weekend, citizens and merchants were delighted, and tourism prospered. The mayor carried the day. This was just one of many novel policies Nîmes launched, as reported in detail by our French FAUI team (Becquart-Leclercq, Hoffmann-Martinot, and Nevers 1987; T. N. Clark, Hoffmann-Martinot, Nevers, and Becquart-Leclercq 1987).

Analysis of the Strategies and Related FAUI Data

This chapter has thus far presented some specifics about major strategies municipal leaders have devised. Next we consider three examples of analyzing the strategies using broader urban theories.

FAUI project participants have studied interrelations among the 33 strategies, using many theories and methods (such as correlation, regression, cluster, and various factor analytic approaches). The best general source is our FAUI project annual, *Research in Urban Policy*, published from 1985 onward.[7] Consider just a few results salient for U.S. urban concerns.

How Do Policy Strategies Cluster? Notes on Both Support for and Contextual Limits of Peterson's Threefold Classification

One theory stressing differences in types of policy strategy is Paul Peterson's *City Limits* (1981). Adapting past work by Musgrave and Lowi, he posited three policy types: (a) developmental—seeking to enhance the property value of the locality, by zoning land, offering tax abatements to developers, building roads and other capital infrastructure facilities, etc.; (b) redistributive—those services that nominally aid the disadvantaged, like public welfare, public health, and public housing; and (c) allocational—including basic municipal services like police and fire and most others not developmental or redistributive. Although some cities may try to pursue each, Peterson held that cities were destined to fail if they attempted redistributive strategies, as they would be undone by market competition with less redistributive cities. Hence, he concluded, the national government is and should be the major provider of redistributive services.

FAUI participants actively debated the theory from its origin. A draft of *City Limits* was used in a seminar Paul Peterson and I taught at the University of Chicago (discussed in alternate weeks with chapters of *City Money*, T. N. Clark and Ferguson 1983). Lorna

Ferguson and Ken Wong were completing Ph.D.s in the seminar. Peterson and Wong later conducted case studies of federal programs, generating a debate by John Kirlin and Robert Stein over the limits of the theory in volume 1 of *Research in Urban Policy* (Peterson and Wong 1985; Kirlin 1985; Stein 1985). In *Research in Urban Policy*, volume 3, Wong (1989) extended the theory in his "political choice" model, essentially arguing that the three policy emphases *may* hold in some cities, nations, and periods, but not others. Wong then edited *Research in Urban Policy*, volume 4, focusing on urban innovations in Chicago (mainly under Mayor Harold Washington), which often included an explicit redistributive component. Did this contradict the theory? At a FAUI roundtable session at the American Political Science Association meeting in Chicago in 1992, many participants reported strong emphasis on redistributive strategies in their locales, often explicitly countering the "antiredistributive" rhetoric of the White House in most of the Reagan-Bush years (1980-1992).

Peterson and Wong illustrated their theories with case studies and fairly simple statistical data. The richer FAUI survey encourages pursuing these issues. The 33 strategies are one set of policy areas surveyed, but do not readily match the three types. The survey also asked mayors and council members of their preferences for spending more, the same, or less in each of 13 areas (items listed in Chapter 2 Appendix). Conducting several simple data reduction procedures using the U.S. FAUI spending data (e.g., factor analysis), we find three clear factors, which correspond neatly to Peterson's three types, a result that should gratify Peterson and others using the typology.

An interesting twist emerges, however, in conducting the same analysis for other countries. The same factors do not emerge. Mayors' policy preferences follow different patterns in countries like France, Norway, Canada, and Australia. In countries with more expansive welfare states, this makes sense for two reasons: welfare, health, hospitals, and housing programs, which are redistributive in the United States, are more broadly middle class and "universal" programs in, for example, most of Europe. Hence they are understandably less statistically distinct from other policy areas (data analysis by Terry Clark, 1992, and Thomas Longoria, 1992). Second, in the United States, "development" is considered a largely private sector policy area, in which the government assists marginally; in France and Norway private and public sectors are often more intertwined. Development is thus a less distinct policy area and more integrated with other functions. Third, insofar as many local governments are less autonomous in retaining distinct local benefits from development inside their boundaries, they may be less concerned with "their" local development than U.S. cities, which compete far more. But these patterns should vary with specific revenue and administrative structures, which differ substantially across countries. Here, as is usually the case, the issue is not that the theory is "right" or "wrong." Rather, comparative analysis makes explicit those contextual conditions (often implicitly assumed or unrecognized) for the theoretical proposition to hold and shows how shifts in these conditions modify the proposition.

Does Fiscal Austerity Generate Urban Innovation?

Our project title—Fiscal Austerity and Urban Innovation—is sometimes construed as implying causality. This too has been debated among project participants. Three major alternatives can join austerity with innovation.

Hypothesis 1: Austerity causes innovation.

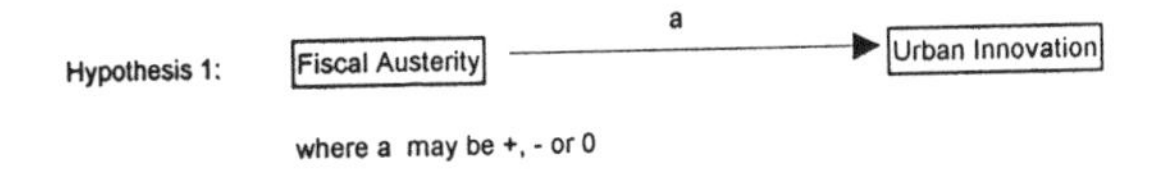

There may be a linkage here, but is it positive or negative? First consider "a" as negative, illustrated by public officials who say "we are too fiscally strained to do new things." This pattern is more likely for innovations that demand a talented professional staff (such as computer programmers), as such staff often leave or get too busy if strain increases. Charles Levine advanced this view and documented several cases of cities that shifted decision-making patterns as fiscal strain rose (Levine, Rubin, and Wolohojian 1981; the authors were honest enough to report that the data often contradicted the theory). Harold Wolman (esp. in Wolman and Davis 1980) suggested a similar but more analytically developed approach, stressing specific stages that cities pass through as their fiscal strain increases, from deferring maintenance to drawing down surpluses to cutting staff, and so on.

Alternatively, "a" may be positive. Crisis can open doors to policy options impossible in "good times" (Wolman 1987). The French "new mayors" who came to office in the early 1980s are graphic illustrations. French bureaucracy, from Napoleon onward, has been a world leader in rigidity (it was often the model for Russians and East Europeans, e.g., Crozier 1964). The French new mayors sought to break dramatically with ideological and bureaucratic tradition, especially of the left, and to provide better services at less cost. They had a huge impact and were widely regarded as successful in cutting through decades of red tape and Red programs. Following their example, even Communist mayors in France started privatizing services—unthinkable just a few years earlier. And French socialists (under President François Mitterrand) abandoned welfare state centralism and implemented major decentralization programs of a sort proposed in the United States by Ronald Reagan, but killed by the U.S. Congress. (See the work of our French FAUI participants, Vincent Hoffmann-Martinot, Jean-Yves Nevers, and Jeanne Becquart-Leclercq, especially Becquart-Leclercq et al. 1987; T. N. Clark et al. 1987; T. N. Clark, Lipset, and Rempel 1993.) Similarly, Scandinavian Social Democrats sought to retain power in the 1980s by cutting back the welfare state they had built in the last half century (Pierre 1993). Dramatic breaks of this sort were also standard in the former Communist countries (T. N. Clark 1993).

Building on these examples, the apparent conflict between a plus or minus can be resolved by adding intervening variables:

Hypothesis 2: Fiscal strain can increase or depress innovation, contingent on intervening factors.

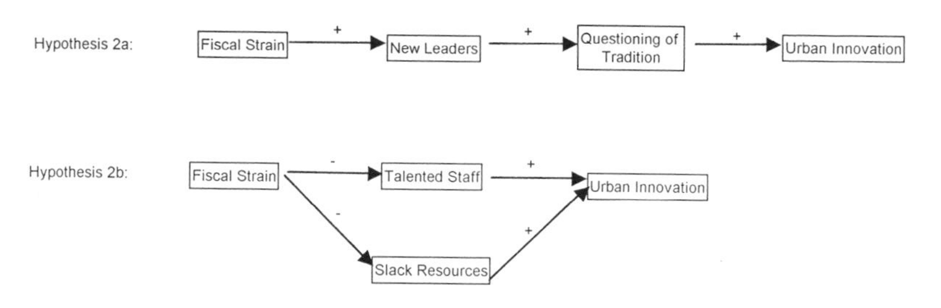

Adding intervening variables this way can often resolve apparently conflicting hypotheses.

A third possibility, however, is that some more general factors generate both fiscal strain and innovation, such that any specific linkage between these two is spurious:

Hypothesis 3: Fiscal strain is only spuriously related to innovation.

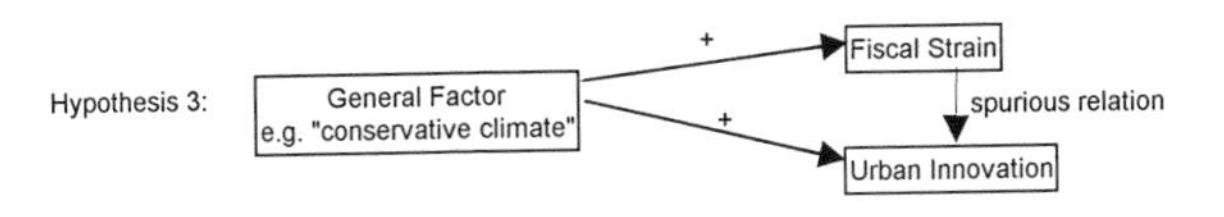

What might such general factors be? People talk of "a conservative climate" such as in post-Proposition 13 California, Margaret Thatcher's Great Britain, or the Chirac government in France. It may be a shift in citizen preferences or a perceived shift that ambitious leaders seek to encourage and articulate. This climate can lead public officials to feel that they "cannot" increase taxes; with inflation this can mean fiscal strain. Does it also mean innovation? Maybe or maybe not. But it clearly implies a different causal pattern.

Many of us have posited one, two, or three of the above patterns using case studies, cross tabulations, or regressions. All are relatively clear. More complex is an interaction model, in which fiscal strain can accentuate or depress other factors. In such interactions, fiscal strain has no independent effect; it only shifts effects of other variables.

Hypothesis 4: Fiscal strain interacts with other factors.

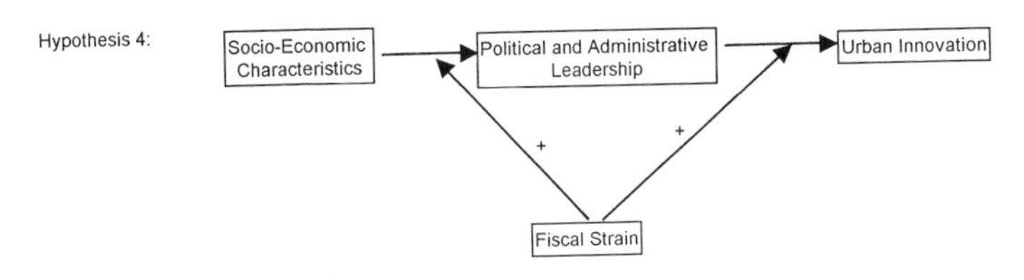

Hypotheses 3 and 4 are partially illustrated by situations in which public officials use the rhetoric of fiscal strain as an "excuse" to justify their actions, saying "we cannot afford" a new program, and the like. Politics and policy are complex enough that one might find support for all four hypotheses in a single city. Nevertheless, the analyst can still clarify past policies and future options by recognizing the quite different causal orderings these each suggest. Mouritzen (1992, esp.

chaps. 4 and 6) and his collaborators provide several useful illustrations of such patterns.

What Is Fiscal Strain?

How have analysts measured fiscal strain to test hypotheses like the four above? T. N. Clark and Ferguson (1983) proposed that fiscal strain is maladaptation by a city to its socioeconomic environment. Strain is measured by ratios of city government revenue, expenditure, and debt divided by private sector resources. Thus a city that loses population but increases spending will increase its strain. A survey of 108 researchers who had published on fiscal strain found a strong majority supporting such a ratio approach (Bingham and Hawkins 1990). Alternatives were funds flow measures (such as deficits) and socioeconomic characteristics alone (such as the "urban distress" measure the U.S. Department of Housing and Urban Development used in the allocation formula for Community Development Block Grants, including components like population change). Correlating these three approaches with political party preference of the researcher showed no statistical differences. Mouritzen (1992) suggested a slightly different approach to fiscal strain. We find the Mouritzen measure conceptually appealing, but it correlates over .9 with grants to city governments. This makes it operationally near identical to grants, which limits its purview of fiscal strain.

What Causes Distinct Strategies, Especially for Productivity?

One answer is a classification that joins strategies with ideologies of leaders, parties, and political cultures. This shows how different specific contexts affect dynamics of leadership and strategy selection. To wit:

1. Traditional right leaders and parties favor less government in general. Of the 33 strategies, they should oppose revenue-enhancing strategies, such as raising taxes, seeking more grants, and so on.

2. Traditional left as well as clientelist/patronage-oriented leaders and parties seek to expand government to foster a wide range of programs. Clientelist leaders are especially concerned that programs generate private or separable goods that they can provide to some but withhold from other constituents. Left leaders should thus support expenditure strategies in general and resist cuts as long and as actively as possible. This in turn should also lead to more revenue-raising strategies, such as increasing taxes, finding new revenue sources, and so on.
3. New Fiscal Populist leaders join the fiscal conservatism of the traditional right with the social progressivism of the traditional left. They reconcile these through productivity strategies—seeking more with the same revenues, improving efficiency in government wherever possible, or at least offering this as their ideological program. Strategies that tap their concerns are the productivity items and others that can encourage efficiency, such as contracting out. (This classification of political cultures and strategies is outlined in T. N. Clark and Ferguson 1983, chap. 10; T. N. Clark 1985a.) Of course these ideological/analytical classifications are subject to considerable adaptation by individual leaders, many of whom do not neatly fit into abstract categories. And almost none explicitly oppose productivity gains. These qualifications should weaken statistical relations among all variables.

What do we find? Mayors of traditional right and left parties generally tend to support or oppose classic welfare state type strategies in some European countries, especially France and Finland (see Table 2.1 in Chapter 2). But in the United States, political parties explain few differences in strategy selection for the left and right. However, New Fiscal Populist mayors differ in supporting productivity strategies. They also support distinct types of management symbolizing efficiency, such as more frequent fiscal reporting, more use of computers, more precise revenue forecasting, and the like. Cities with these sophisticated management techniques are also more likely to adopt productivity improvement strategies. These patterns have been analyzed in a slightly different manner by seven different teams of U.S. researchers, summarized in Table 8.2 and in Appleton and T. N. Clark (1989). Cities with New Fiscal Populist leaders were more likely to choose productivity strategies in three of four studies that examined them and in some models estimated in the fourth study. Strategies thus vary by context, although these are moderate, nondeterministic patterns.

TABLE 8.2 New Fiscal Populist Leaders and Sophisticated Staff More Often Use Productivity Strategies in Seven Past Studies From the U.S. Fiscal Austerity and Urban Innovation Project

	CBL	*CW1*	*CW2*	*B*	*MP*	*A*	*H*
R		.36	.49				.50
R^2	.14			.13	.13	NA	.25
R^2 Adj					.09		
Independent Variables:							
Fiscal Strain							
Expend/PCIncome	0	0	0	0			
Chng in Rev 75-82				.14		Low	
Chng in debt 75-80					–.12	Low	
Political Culture							
Democratic	0			.09		Varies	
Republican	0			0		by	
Ethnic	–.20		0	0		Period	
New Fiscal Populist	.16		.14	.13		+ and –	
Mayor							
Democratic Party	–.17			–.10			
Fiscal Liberalism			0		0		
Fisc Lib*Mys Power	.16			0			
Organized Groups							
Active Lib Grps			.16		0		
Active Demo Party				–.12			
Salient Problems							
All					0		
Federal Cuts						.14	.17
State Cuts		0	0			.18	–.26
Inflation etc.						.13	
Decline/Unempl		0	0			.17	–.03
Municipal Employees		.15				.09	.39
Taxpayers			.26			.18	.39
Socioecon Chars							
City Size		0	0				
Population Change			0				
Per Capita Income		.02	0	0			–.03
Foreign Stock	0			0			
Percent Black	0			0			
Percent Minority			0				
Percent Poor	0						
Manuf, Age, & Region					.23		

continued

TABLE 8.2 Continued

	CBL	*CW1*	*CW2*	*B*	*MP*	*A*	*H*
Administrative Staff							
City Manager Govt	0	0	0	0	0		
Fiscal Mgmt Sophis.		.27	.25		.21	High	
Staff Power	.26		0	.29			
Other							
January Temp.			0				

SOURCES: CBL = T. N. Clark, Burg, and Landa (1984).
CW1 = C. Clark and Walter (1986); dependent var is "professional retrenchment" index summing 15 strategies.
CW2 = C. Clark and Walter (1987).
B = Burg (1986), a refinement of CBL.
MP = Morgan and Pammer (1986); cf. also Pammer (1986) for more detail.
A = Appleton (1989); reports *r*s with various controls included; distinguishes strategies by year; NFP political culture increases productivity in early years, opposite later.
H = Hawkins (1989); most variables are indexes of 4-7 components, sometimes reported twice in this table if two components in one index, e.g., municipal employee and taxpayer problems; higher R^2 may be due to lower *N*.
NOTE: Most studies classified the 33 strategies into the same three types (revenue, expenditure, productivity). Most used the same independent variables for all three types of strategies, although this table reports results only for the number of productivity strategies adopted by the city. All studies (except A) use OLS regression. The R, R^2, and Adjusted R^2 at the top are for the full equation as reported in each study. Coefficients below them are betas, reported only if greater than .10 significance, otherwise shown as 0 if included in the model or left blank if not included. For example, the CBL study used 12 independent variables generating an R^2 of .14; the beta for its fiscal strain measure was 0. Some studies combine items into indexes; the main item is then listed. A and H are distinctive enough in method from others that readers should consult the original sources.

Conclusion

This chapter has outlined specific strategies U.S. cities have used to adapt to fiscal stress during the last two decades. The policy-oriented reader can find here strategies that work. The chapter showcases strategies "in place" in actual cities, rather than listing our personal recommendations. Many cases are briefly summarized to indicate specific choices that local officials have made, often under considerable duress in these years of general austerity. Cases include, for instance, joint agreements to form a tree-buying consortium; contracting out for computer acquisition, park construction, water meter inspection, and refuse removal; and using a "shared savings" method to reduce energy costs with zero investment by the city.

The chapter then raises the more abstract question "Where do different strategies emerge?" General pattern: diversity across cities; innovations surface in many locales. An intriguing policy implication of this loose coupling is that innovation is minimally predicted by wealth or fiscal stress. We stress this "nonresult" with wealth, as one so often hears the opposite: "Poor cities cannot afford to do this or that . . ." But our "nonresult" implies a distinct policy implication. If more affluent cities are often no different from others in innovation and strategy selection, staff and leaders in both rich and poor cities can develop new and unusual patterns for service delivery to their constituents. Policy implication: less affluent cities should not be considered by their leaders, citizens, or outsiders as "unable" to innovate using many of the strategies discussed. Dramatic innovations are often feasible with minimal time and minimal funding and by minimal staff. Top staff encouragement is far more critical than having some affluent residents.[8]

Although we do find some distinct patterns of policy strategy by type of city and leader, we stress that practically all relations are weak in the United States. Policy implications: U.S. local leaders are minimally constrained by party ideology; Democrats are just as likely as Republican mayors to pursue many strategies—in contrast to European countries where parties are far more structured. Chapter 2 showed certain hierarchical and racial cleavages that make U.S. cities resemble European party patterns. Nevertheless, that chapter and results in this one indicate loose linkages among all variables. Policy implication: U.S. local leaders are quite weakly constrained by their fiscal strain and staff size, their political parties, their residents' wealth and race, and most other factors often seen as "discouraging innovation." U.S. leaders lead.

Notes

1. Some of the more useful work on innovation emerged from a National Science Foundation Program in the area. See Bingham (1976), Wolman (1987).

2. E.g., the Advisory Commission on Intergovernmental Relations advocated higher property taxes for years, arguing that they were progressive. Harberger's analysis is complex, but turns on including all capital of individuals (not just private homes) and including individuals as shareholders who own stock in corporations in other cities. This implies assigning property taxes paid by businesses nationally to individuals, as individuals own businesses. When these more comprehensive calculations are made,

property taxes emerge as heavily taxing the more affluent. The issues are summarized clearly in Aaron (1975). An assessment of alternative city revenue sources and property taxes is in a report commissioned by the City of Chicago to help it find new revenues (T. N. Clark 1985b). Criteria discussed there include fairness, equity, ability to pay, benefits received, neutrality, economic development impacts, elasticity, diversification, and feasibility.

3. From 1978 to 1985, federal grants fell $109 per capita for all California cities. Property taxes fell $47 per capita. User fees made up most of the difference, rising $120 per capita. Total per capita revenues for the average California city in 1985 were $612. Data computed from California State Comptroller by Gary Reid (1988).

4. Miranda (1992) has elaborated this point.

5. Then president of the Municipal Finance Officers Association and author of dozens of books on how to manage city finances properly, Moak was pressed by Mayor Rizzo to come up with more money without raising taxes or visible revenues. He was sophisticated enough to implement accounting "innovations" that ostensibly raised revenues, such as changing the fiscal year by one month. Hoping to sell bonds to buyers who were boycotting Philadelphia, he advertised in the *Wall Street Journal* that the city had a large surplus. But a few weeks later, news spread that this was a "manufactured surplus," and it became a national scandal. Pursuing these practices near the time of the New York fiscal crisis led Moak, with New York Mayor (and former Comptroller) Abraham Beame to become the two major local officials investigated for possible fraud by the Securities and Exchange Commission. Moak tragically died of a heart attack. He dramatically illustrates how even the most professionally concerned staff can pursue questionable practices under political pressure (cf. T. N. Clark 1981, pp. 150-151).

6. Illinois has more incentives to cooperation because it has more local governments than any other U.S. state—not just counties and municipalities, but thousands of special purpose governments for schools, roads, public buildings, etc.

7. But for more specifics, see, on several European countries: Clarke (1989) and Mouritzen (1992); on Texas counties: Vertz (1985); on New Jersey medium-sized and small towns: Friedman and Rhomberg (1988); on Wisconsin and Florida: Magill (1988, 1990); on France: T. N. Clark, Hoffmann-Martinot, Nevers, and Becquart-Leclercq (1987), and Bernier (forthcoming); on the Netherlands: Spit (1993); on Poland: Bartowski, Kowalczyk, and Swianiewicz (1990); on Norway: Baldersheim (1990).

8. We have conducted case studies of some dramatic cases that illustrate innovation by cities with poor residents; to illustrate specifics, Waukegan, Illinois, is one striking example (Walzer 1985; T. N. Clark forthcoming).

References

Aaron, Henry. 1975. *Who Pays the Property Tax?* Washington, DC: Brookings Institution.

Appleton, Lynn M. 1989. "Determinants of Innovation in Urban Fiscal Strategies." Pp. 51-72 in *Decision on Urban Dollars.* Vol. 3, *Research in Urban Policy,* edited by T. N. Clark, W. Lyons, and M. Fitzgerald. Greenwich, CT: JAI.

——— and Terry Nichols Clark. 1989. "Coping in American Cities: Fiscal Austerity and Urban Innovation in the 1980s." Pp. 31-58 in *Urban Innovation and Autonomy:*

The Political Implications of Policy Change. Vol. 1, Sage Series on Urban Innovation, edited by Susan Clarke. Newbury Park, CA: Sage.

Baldersheim, Harald, ed. 1990. *Ledelse og Innovasjon i Kommunene*. Bergen: LOS Center.

———. 1992. "Fiscal Stress and Local Political Environments." Pp. 83-100 in *Managing Cities in Austerity*. Vol. 2, Sage Series on Urban Innovation, edited by Poul-Erik Mouritzen. London: Sage.

Bartowski, Jerzy, Andrzej Kowalczyk, and Pawel Swianiewicz. 1990. *Strategie wladz lokalnych*. Warsaw: Institute of Space Economy, University of Warsaw.

Becquart-Leclercq, Jeanne, Vincent Hoffmann-Martinot, and Jean-Yves Nevers. 1987. *Austerite et Innovation Locale, Les Strategies Politico-Financieres des Municipalites Urbaines Dans La Crise*, 2 vols. Bordeaux: Cervel.

Bernier, Lynne Louise. Forthcoming. *Fiscal Austerity and Urban Innovation in France* (tent. title).

Bingham, Richard. 1976. *The Adoption of Innovation by Local Government*. Lexington, MA: D. C. Heath.

——— and Brett W. Hawkins. 1990. "A Test of Political Bias in Scholars Preference for Measuring Fiscal Strain." *Urban Affairs Quarterly* 25,3 (March):515-23.

Brooks, Stephen C. 1993. "Urban Fiscal Stress: A Decade of Difference." Presented to 1993 annual meeting of Midwest Political Science Association, April 15-17, Chicago.

Burg, Margaret H. 1986. "The Effect of Political Culture on Urban Financial Decisions." M.A. essay, Dept of Political Science, University of Chicago.

Clark, Cal and B. Oliver Walter. 1986. "City Fiscal Strategies." Pp. 89-115 in Vol 2B, *Research in Urban Policy*, edited by T. N. Clark. Greenwich, CT: JAI Press.

——— and B. Oliver Walter. 1987. "Political Culture, Management Style and Fiscal Austerity Strategies." Presented to Midwest Political Science Association, Chicago.

Clark, Terry Nichols. 1981. "Fiscal Strain and American Cities." Pp. 137-56 in *Urban Political Economy*, edited by Kenneth Newton. London: Frances Pinter.

———. 1985a. "Choose Austerity Strategies That Work for You." Pp. 71-88 in *Coping With Urban Austerity*. Vol. 1, *Research in Urban Policy*, edited by T. N. Clark. Greenwich, CT: JAI.

———. 1985b. *Revenues for Chicago*. City of Chicago, White Paper 1.

———. 1988. "Costless Computers? Almost—How to Install a Computer System at No Cost to the Taxpayer—Lessons for County and Municipal Governments From Cook County, Illinois." A Case Study Based on Interviews with Henry "Bus" Yourell, Cook County Recorder, Paul Cericola, Recorder's Office, Bob Dickey, Greg English, Larry Tonelli, and John E. Lockwood, Business Records Corporation. Portions published in Terry Nichols Clark, "Cook County's Cost Free Computer System." *Illinois Government Finance Review* (July 1988): 9-11.

———, ed. 1990. *Monitoring Local Governments: How Personal Computers Can Help Systematize Municipal Fiscal Analysis*. Dubuque, IA: Kendall-Hunt.

———. 1993. "Local Democracy and Innovation in Eastern Europe." *Government and Policy* 11:171-98.

———. Forthcoming. "Clientelism, USA." In *Democracy, Clientelism, and Civil Society*, edited by Luis Roniger and Aysa Gunes-Ayata. Boulder, CO: Lynne Rienner.

———, Margaret H. Burg, and Martha Diaz Villegas de Landa. 1984. "Urban Political Cultures and Fiscal Austerity Strategies." Presented to annual meeting of American Political Science Association, Washington, DC.

———, Daniel K. Crane, Ann L. Kelley, Joanne Malinowski, Melissa Pappas, Gregory Wass. 1989. *Taxes in Chicago and Its Suburbs*. City of Chicago, Office of Comptroller, White Paper 3. (Abbreviated version published in *Politics of Policy Innovation in Chicago*.) Vol. 4, *Research in Urban Policy*, 1992, edited by Kenneth K. Wong and Terry Nichols Clark, pp. 3-30. Greenwich, CT: JAI.

———, G. Edward DeSeve, and J. Chester Johnson. 1985. *Financial Handbook for Mayors and City Managers*, 2nd ed. New York: Van Nostrand Reinhold.

——— and Lorna Crowley Ferguson. 1983. *City Money: Political Processes, Fiscal Strain, and Retrenchment*. New York: Columbia University Press.

———, Vincent Hoffmann-Martinot, Jean-Yves Nevers, and Jeanne Becquart-Leclercq. 1987. *L'innovation municipale a l'epreuve de l'austerite budgetaire*. Bordeaux, France: Report to the Plan Urban from CERVEL, IEP de Bordeaux.

———, Seymour Martin Lipset, and Michael Rempel. 1993. "The Declining Political Significance of Social Class." *International Sociology* 8(3):293-316.

Clarke, Susan, ed. 1989. *Urban Innovation and Autonomy: The Political Implications of Policy Change*. Vol. 3, Sage Series on Urban Innovation. Newbury Park, CA: Sage.

Crozier, Michel. 1964. *The Bureaucratic Phenomenon*. Chicago: University of Chicago Press.

Ferguson, Lorna C. 1993. "Building Public Infrastructure With Private Money." Presented to Fiscal Austerity and Urban Innovation Project Tenth Anniversary Conference, Tampere, Finland.

Friedman, Judith J. and Christopher Rhomberg. 1988. "Fiscal Conditions in New Jersey Municipalities." Department of Human Ecology, Rutgers University.

Hawkins, Brett W. 1989. "A Comparison of Local and External Influences on City Strategies for Coping With Fiscal Strain." Pp. 73-92 in *Decision on Urban Dollars*. Vol. 3, *Research in Urban Policy*, edited by T. N. Clark, W. Lyons, and M. Fitzgerald. Greenwich, CT: JAI.

Im, Angela. 1987. "Innovative City Management in Downers Grove." Chicago: Urban Innovation in Illinois, Research Report 196.

Jeffe, Douglas and Sherry B. Jeffe. 1988. "Proposition 13 Ten Years Later." *Public Opinion* 11(May/June):18-19.

Kirlin, John. 1985. "Comparing Federal Education and Housing Programs." Pp. 349-56 in *Coping With Urban Austerity*. Vol. 1, *Research in Urban Policy*, edited by T. N. Clark. Greenwich, CT: JAI.

Levine, C. H., I. S. Rubin, and G. C. Wolohojian. 1981. *The Politics of Retrenchment*. Beverly Hills, CA: Sage.

Longoria, Thomas. 1992. "Comparative Analysis for the City Limits Typology." Dept. of Political Science, Texas A&M University.

MacManus, Susan A. 1983. "State Government: The Overseer of Municipal Finance." Pp. 145-184 in *The Municipal Money Chase: The Politics of Local Government Finance*. edited by A. M. Sbragia. Boulder, CO: Westview.

Magill, Robert S. 1988. "Urban Fiscal Strain and Regionalism: The Case of Wisconsin and Florida." *Wisconsin Sociologist* 25,4 (Fall):131-38.

———. 1990. "Community Values, Regional Differentiation and Social Welfare." *Wisconsin Sociologist* 27, 1(Winter):15-21.

Malme, R. Undated. *Planning a Mining and Reclamation Project in an Urban Environment*. Elgin, IL: City of Elgin.

Miranda, Rowan. 1992. "Privatizing City Government." Ph.D. dissertation, Irving B. Harris School of Public Policy Studies, University of Chicago.

Mouritzen, Poul-Erik, ed. 1992. *Managing Cities in Austerity.* Vol. 2, Sage Urban Innovation Series. London: Sage.

Mushkin, S. J. and C. L. Vehorn. 1980. "User Fees and Charges." Pp. 222-234 in *Managing Fiscal Stress,* edited by C. H. Levine. Chatham, NJ: Chatham House.

Pagano, Michael A. 1993. "Balancing Cities' Books in 1992." *Public Budgeting and Finance* 13, 1(Spring): 19-39.

Peterson, Paul E. 1981. *City Limits.* Chicago: University of Chicago Press.

——— and Kenneth Wong. 1985. "Toward a Differentiated Theory of Federalism." Pp. 301-24 in *Coping With Urban Austerity.* Vol. 1, *Research in Urban Policy,* edited by T. N. Clark. Greenwich, CT: JAI.

Pierre, Jon. 1993. "Legitimacy, Institutional Change, and the Politics of Public Administration in Sweden." *International Political Science Review* 14:387-402.

Reid, Gary J. 1988. "California Cities and Proposition 13." *Public Budgeting and Finance* 8, 1(Spring):30-37.

Rigos, P. N. 1986. "The Role of User Fees in Municipal Budgets: Fiscal Health Tool or Spending Device." Paper presented at the Southwest Political Science Association meetings, San Antonio, Texas.

Spit, Tejo. 1993. *Strangled in Structures: An Institutional Analysis of Innovative Policy by Dutch Municipalities.* Faculty of Geographical Sciences, University of Utrecht.

Stein, Robert M. 1985. "Implementation of Federal Policy." Pp. 341-47 in *Coping With Urban Austerity.* Vol. 1, *Research in Urban Policy,* edited by T. N. Clark. Greenwich, CT: JAI.

Urban Innovation in Illinois. 1988. "Urban Innovations: The Winners." *Illinois Government Finance Review* (July):2-5.

———. 1989. *Urban Innovation: Who's Doing It? What Really Works?* Chicago: Urban Innovation in Illinois.

Vertz, Laura L. 1985. "County Governments in Texas." Dept of Political Science, North Texas State University; presented to 1985 annual meeting of the Southwestern Social Science Association, Houston.

Walzer, Norman. 1985. "Fiscal Austerity in Mid-Sized Cities." Pp. 161-74 in *Coping With Urban Austerity.* Vol. 1, *Research in Urban Policy,* edited by T. N. Clark. Greenwich, CT: JAI.

——— and Steven C. Deller. 1993. "Federal Aid and Rural County Highway Spending." *Policy Studies Journal* 21(2):309-24.

Wolman, Harold. 1987. "Innovation in Local Government and Fiscal Austerity." *Journal of Public Policy* 6(2):159-80.

——— and Barbara Davis. 1980. "Local Government Strategies to Cope With Fiscal Pressure." Pp. 231-48 in *Fiscal Stress and Public Policy,* edited by Charles H. Levine and Irene Rubin. Beverly Hills, CA: Sage.

Wong, Kenneth. 1989. "Toward a 'Political Choice' Model in Local Policy Making." Pp. 217-45 in *Decision on Urban Dollars.* Vol. 3, *Research in Urban Policy,* edited by T. N. Clark, W. Lyons, and M. Fitzgerald. Greenwich, CT: JAI.

Technical Appendix

Terry Nichols Clark

The FAUI Project. The Series Introduction provides a general overview not repeated here.

Survey Design and Data Collection Procedures. The population surveyed was all U.S. municipalities with more than 25,000 residents in 1980. Municipalities are termed city, village, or town in different states; we often use "city" for brevity. Municipalities are general-purpose local governments responsible for a wide range of functions. The range is measured by our Functional Performance (FP) Index, below. Questionnaires were mailed to three officials in each municipality: mayor, chief administrative officer (or city manager), and a leading council member (often chair of the city council finance committee). The council member to contact was usually identified by a phone interview with the city clerk. Many survey items were adapted from recommendations of specialists who summarized hypotheses from their respective subfields of urban policy and proposed items for a common survey (see Clark 1981). Questionnaire items were prepared by Terry Clark, Richard Bingham, and Brett Hawkins and then improved following pretests and comments from many persons who helped conduct the survey. Questionnaires were mailed by 26 teams (listed in the Series Introduction) to local officials in their geographic areas. Most teams repeated two or three mailings to nonrespondents, and then conducted two or more telephone follow ups. The approach included many hints and specifics from Dillman (1979), Bradburn and Sudman (1979), and past NORC surveys using the Permanent Community Sample (see Clark and Ferguson 1983, Appendix 1). Response rates were about 45 percent, similar to past surveys of local officials (see Caputo and Cole 1977). A few variables came from surveys by the International City Management Association (ICMA), which had about 40 percent response rates. Other data were merged from U.S. Censuses of all cities over 25,000 in 1970 and 1980 (N = about 950 cities each year). Some 100 cities exceeded 25,000 in 1970, but fell below 25,000 in 1980, or vice versa. We retained all cities with a population greater than 25,000 in either year; the N thus rose to 1,030.

Sample Selection Bias Assessment, Jackknifing, and More

Merging data from different sources generates different *N*s of cities for different variables. This raises the question of whether results are distorted when we include certain cities with data for a specific variable, but omit other cities lacking data for the same variable. Bias introduced by such selective nonresponse has generated considerable methodological research in recent years. Because some of our analyses combine several variables (in multiple regressions, etc.) generating different subsets of cases, we sought to assess possible bias introduced by using different subsets of cases. Our concerns were thus twofold: (a) sample selection bias and (b) selecting for analysis subsets of cases that might differ from other subsets in substantively meaningful ways. We assessed both types of potential bias in two principal ways. First we used variations of the Heckman et al. hazard rate, discussed below. But simultaneous with exploring the hazard rate approach, we developed a different method, discussed next.

The Subset Comparison Method. The subset comparison method is a set of interrelated procedures we devised that combine testing for sample selection bias with intentional jackknifing across subsamples. It is statistically austere, but more comprehensive in scope of variables and subsets of cases considered than the more aggregated Heckman et al. hazard rates. It generates a map of possible response bias patterns and details variations across subsets of cases. Others might find the method helpful, as it complements the hazard approach used by Heckman et al. in identifying potentially troublesome variables and subsets of cases that can be pursued with other tools.

TABLE A.1 Illustration of Method for Selecting 15 Subsets of Cases

Total N of Cases for Each of Six Major Subsets	N	Number of Cases Found by Cross-Tabulating Six Variables With One Another, Generating 15 Subsets of Cases Used in Analyses					
		MAYHIDUM	CAOHIDUM	STRTHID	STRTLOD	V7NOSDUM	COUNRDUM
MAYHIDUM	344						
CAOHIDUM	481	236					
STRTHID	500	251	469				
STRTLOD	279	131	275	278			
V7NOSDUM	176	176	118	125	63		
COUNRDUM	280	143	187	201	106	74	

SOURCE: U.S. Fiscal Austerity and Urban Innovation Project surveys.

It goes as follows. We selected several subsets of cities, including officials responding to particular items, or combinations of items, with (a) high and (b) low response rates. Nonresponses were generated because mayors, chief administrators, and council members responded at different rates. Further, responding officials still left certain items blank, some more than others. The majority of nonresponses occurred on items that asked the respondent to answer, for instance, about his or her spending preferences for those specific items performed by his or her municipal government. Thus about two thirds of local officials did not respond to items dealing with public health and welfare, because these are not municipal functions in most jurisdictions, but are performed by another local government, such as a county. See "Functional Performance Analysis" below. Thus here, as in many cases we examined closely, nonresponse is less a question of bias than of substantive differences across subsets of cases. Such differences deserve substantive analysis and interpretation, not just waving a methodological red flag.

We chose six variables to help map varying response patterns. Three were chosen for the highest response rates by the mayor, administrator, and council member ($N = 500$ to 344). Three others had the lowest response rates by the same three officials ($N = 280$ to 176). Table A.1 shows the response N for the six variables (down the first column) and the Ns for the 15 bivariate pairwise combinations of these variables (Ns vary from 469 to 74).

Next we selected 40 variables that met one or more of the following criteria. The variables selected were (a) widely used Census measures which may be associated with response bias (e.g., city level income and ethnicity), (b) central variables in our substantive analyses, and (c) ones which varied considerably in *N*, from population size (*N* = 992) to an index of mayor's fiscal liberalism (V7NOSME, *N* = 176). We computed means and standard deviations for these 40 variables for each of the 15 subsets of cities, listed in Table A.2, generating *N*s from 992 down to 29. We arrayed the data in spreadsheet form to ask: How much did the 40 rows of means and 40 rows of standard deviations differ across the 15 subsets (boxes/columns) of cities? The main comparison was between (a) the first two columns (both in Box 1 of Table A.2) of means and standard deviations for the highest *N*s available for all responding cities, and (b) each of the 15 subsets of cities, first those with higher and then those with lower *N*s. Comparisons were conducted by calculating a *subset ratio:*

$$R_{sv} = S_{sv}/S_{tv} \qquad (1)$$

where

R_{sv} = Subset ratio, computed for each of 15 subsets of cities (1...15) for each of 40 variables (1...40)

S_{sv} = Statistic for city subset s on variable v, where for each v two statistics (S) are calculated:

$$S = \begin{cases} 1 = \text{Mean and} \\ 2 = \text{Standard deviation} \end{cases}$$

S_{tv} = Same as S_{sv} except that all calculations are for the total set of cities for which data are available for variable v.

To illustrate, the mean for Variable 1 (per capita income) for City Subset 1 was $7,956, and $7,655 for total set of cities. Applying the subset ratio formula in Equation 1:

$$R_{11} = 7965/7655 = 1.040496$$

This result is typical: almost all means and standard deviations differed minimally. We were frankly surprised that so few differences emerged, even with some *N*s as low as 29.[1] In particular, we

TABLE A.2 Three Sets of Cases and 40 Variables Illustrating Data Matrices

Variable Name	Variable Acronym	Box 1. Entire Population			Subset 1. Select if MAYHIDUM EQ 1 and CAOHIDUM EQ 1			Subset 15. Select if V7NOSDUM EQ1 and COUNRDUM EQ1		
		Mean	Std	N	Mean	Std Dev	N	Mean	Std	N
PER CAPITA INCOME 1980	V904	7655.58	1905.53	936	7956.33	1954.48	230	7288.32	1375.83	72
POPULATION CHANGE 70	POPCHG	0.29	2.28	939	0.27	0.8	230	0.1	0.35	72
POPULATION 1980 FROM	POP80	96809.68	290993	992	149368.29	526983	236	228116.6	848109.4	74
RATIO POP86 OVER POP	POP8086	1.01	0.04	936	1.02	0.04	229	1.01	0.05	69
POP BLACK 1980,CALC	BLK80	0.11	0.16	935	0.1	0.13	229	0.12	0.12	72
POP HISPANIC 1980,CA	HIS80	0.07	0.12	935	0.05	0.08	229	0.04	0.05	72
FOREIGN STOCK,CODED/	FORNSTK	7.17	7.16	998	6.53	5.62	233	5.53	4.96	72
%W/ED25&OV W/4YR HS-	PV870	0.7	0.12	956	0.72	0.11	234	0.7	0.1	73
% W/ED 4YRS COLEGE (	PV871	0.18	0.1	956	0.2	0.1	234	0.19	0.1	73
FORM OF GOVERNMENT H	FOG	1.66	0.55	1001	1.67	0.51	235	1.49	0.53	74
FUNCTIONAL PERFORMAN	FPIX84NC	422.9	200.85	988	408.07	178.35	235	418.72	197.89	73
Ratio 85/80 General	GX8580	2.14	13.76	924	2.14	8.15	223	1.5	0.54	68
CHANGE IN GEN REV SH	GRS8580	1.04	0.33	893	1.03	0.31	212	0.99	0.25	66
Change in FED IG Rev	FIGR8580	1.3	1.58	895	1.32	1.76	212	1.42	1.53	66
Common Functions Cha	CF8580	1.83	5.92	923	2.1	7.85	223	1.52	0.59	68
Mayrs Party 1= Rep, 2 = Dem	PARTMAY	1.58	0.46	324	1.57	0.46	221	1.67	0.42	67
MYRS PARTY,1=REP,2=D	SPARTMAY	1.59	0.49	283	1.58	0.49	192	1.71	0.46	55
MY SP PREF, RECODED	V7RS	3.11	0.77	344	3.12	0.75	236	3.16	1.11	74
FISLIB W/OSAME SP,RE	V7NOSME	3.22	1.07	176	3.24	1.05	118	3.16	1.11	74
Q21 OWN BUDGET POSTI	V133R	1.31	0.39	395	1.29	0.4	236	1.39	0.44	74
Q21 OWN BUDGT POSTN,	V133RS	1.24	0.43	293	1.25	0.43	194	1.37	0.49	60
Q31 PREF RE LOCAL TA	V143R	2.93	0.82	395	2.9	0.86	236	2.87	1	74
Q31 LOC TAX BURD LO=	V143RS	2.37	0.8	334	2.41	0.79	225	2.46	0.93	69
My-Soc Cons Index(0+	SOCCONS	1.51	0.26	389	1.49	0.27	234	1.5	0.27	72
Q16 MY BUY GUN? RECO	V128RS	1.32	0.47	294	1.31	0.46	191	1.3	0.46	61
Q16 POLICE PERMIT TO	V128R	1.36	0.41	389	1.34	0.42	234	1.33	0.43	72
Q17 MY BUSING, RECOD	V129RS	1.83	0.38	262	1.82	0.38	169	1.76	0.43	54
Q17 STAND ON BUSING	V129R	1.72	0.35	389	1.73	0.36	234	1.69	0.39	72
Q19 SEX ED, HI = AGA	V131RS	1.13	0.34	280	1.11	0.32	187	1.15	0.36	60
Q19 SEX ED IN PUBLIC	V131R	1.23	0.33	389	1.19	0.32	234	1.21	0.35	72
Q20 ABORTION, HI = A	V132RS	1.89	0.53	285	1.88	0.52	190	1.95	0.5	61
Q20 ABORTION, HI = A	V132RS	1.89	0.53	285	1.88	0.52	190	1.95	0.5	61
Q30 TAKE POSITION UN	V142R	2.48	0.79	395	2.45	0.79	236	2.51	0.91	74
Q30 TAKE POS UNPOP W	V142RS	2.37	0.8	334	2.41	0.79	225	2.46	0.93	69
FISCAL MGMT IX=(246R	FMIX	3.44	0.73	450	3.49	0.74	214	3.45	0.81	47
Q8.1 SEEK NEW LOCAL	V228R	2.91	1.65	517	3	1.62	236	2.84	1.59	59
Q8.2 GET ADDITIONAL	V229R	2	1.54	517	1.96	1.42	236	1.84	1.43	59
Q8.1 SEEK NEW LCL RE	V228RS	3.77	0.94	379	3.78	0.97	178	3.61	0.99	44
Q8.2 GET ADDL IGR,RE	V229RS	3.26	0.93	279	3.1	0.86	129	3.17	0.76	29
RESPONSIVENSS TO COU	V105RS	3.95	0.85	290	3.9	0.89	199	3.95	0.85	57
Q23A MOST PARTI,CM-P	V421S	3.19	1.53	280	3.34	1.53	109	3.19	1.56	74

NOTE: This shows only the first and last of the 15 subsets of cities that were contrasted with the total population (first box). Subset 1 has a relatively large number of cases; subset 15 a small number.

found few differences in subset ratios for cities lower in income, with more blacks, or with Democratic mayors. The one variable most associated with subsample variations was population size. Like all other variables analyzed in this sample selection assessment, it was used unlogged, although it was logged in many regression analyses. Logging a skewed variable like population size reduces its subset ratio. Using unlogged variables, as we do here, thus should tend to increase subset ratios for skewed variables. Nevertheless, apart from population size, the 39 other variables showed minimal subset ratios. To assess patterns across subsets of cities, we counted the number of ratios below .33 or .66, and above 1.5. We calculated 840 R_{svs}, that is,

$$N_{R_{sv}} = 2 \text{ (Statistics = 1 mean + 1 standard deviation)} \quad (2)$$
$$\times 15 \text{ (s = Subsets of cities)}$$
$$\times 40 \text{ (v = Variables)}$$
$$= 840$$

Of these 840 ratios, 69 were less than .33, 30 less than .66, and just 2 exceeded 1.5. The bulk of the high and low ratios came from a single variable: population size. For instance, 23 of the 30 ratios below .66 were for population size; only 7 of 840 other ratios ever differed this much. In sum: 3 percent of the ratios were below .33, but less than 1 percent were below .66 or exceeded 1.5. If population size is excluded, less than 1 percent of the ratios were above or below these limits (see Table A.3). Following general practice for work on such sampling issues, we do not calculate probabilities for the sampling ratios of means and standard deviations (e.g., Kish 1965; Blau and Duncan 1967, pp. 471ff).

These results indicate remarkable stability across both means and standard deviations for even those combinations of variables a priori most likely to generate the most acute sample selection biases.

The subset comparison analysis we have developed here goes further than sample selection as it incorporates the logic of jackknifing, that is, partitioning the sample into subsets and examining key findings for consistency across subsets of cases. However, unlike common jackknifing that partitions a sample into arbitrary subsets of cases (e.g., split halves or 10 deciles), it is purposive jackknifing which creates partitions seeking to *maximize* the potential for response bias. It also encourages identifying subsets of cases that may systematically differ for substantively important reasons that demand substantive interpretation (for example, population size) that should not be dismissed as sampling error, but considered further for interpretation.

TABLE A.3 Number and Percent of Cases With Test Ratios Exceeding Three Different Levels

	Three Levels: Cutting Points = .33, .66 and 1.5					
	>.33		>.66		>1.5	
Variables	N	%	N	%	N	%
Population in 1980	39	4.64%	23	2.74%	2	0.24%
Other 39 Variables	39	3.57	7	0.83%	0	0.00%
Total Variables with High Test Ratio Scores	69	8.21	30	3.57%	2	0.24%
Total N of Cases Considered	840		840		840	

Comparisons of Correlations. The subset comparison method is readily extended to statistics besides means and standard deviations. We pursued possible differences in *relations* among variables that might emerge from subsample differences in several ways. One was to compare several dozen Pearson *r*s for bivariate relations between the same two variables for separate subsets of cities, such as the 15 subsets above. For example, we correlated population size with per capita income, first for the total set of cities, and second for that subset of cities where both the mayor and council member responded, and so forth. If both *r*s were similar, then one may be less concerned about distinct patterns of results in subsets of cases. (This is broadly analogous to comparisons with a multimethod-multitrait matrix.) The great bulk of the *r*s were similar for all subsets of cities. Occasional *r*s differed; they were sometimes lower for subsets with lower *N*s. But this analysis also illustrates the limits of the Pearson *r*: differences in *r*s may derive from differences in the amount of variance rather than in slopes, but slopes are more important for most of our analyses. We thus do not summarize these results systematically. Still, if one finds such minimal differences, as we did, this is positive evidence of consistency across subsamples.

Hazard Rates. We computed several hazard rates using the general method of Heckman (1979) and others. Following the Olsen (1980) and Achen (1986) extension of the method, we generally used OLS rather than logit estimation, computing first a hazard rate variable, and second, including it as a predictor variable with others in subsequent multiple regressions.

The hazard rate approach starts from two equations:

$$Y_1 = \mathbf{X}_1 b1 + \mathbf{U}_1 \tag{3}$$

$$Y_2 = \mathbf{X}_2 b2 + \mathbf{U}_2 \tag{4}$$

where **X**s are vectors of exogenous variables, **U**s error terms, and the b's coefficients predicting the Ys. Equation 3 is the "substantive equation" that one seeks to solve, but in a situation where sample selection limits the range of variables, only Equation 4, the "selection equation," can be estimated. If there is "bias" in selection, the two error terms **U** will not have the normally assumed statistical properties. But if one can theoretically state the specific pattern of "bias," and how it may be modeled, then Heckman reasons it may be possible to identify its magnitude in empirical work and at least provide the researcher with a flag to be careful. How? By introducing a "hazard coefficient." Define T where Y1 is observed if Y2 > T, and Y1 is not otherwise observed. Then Y3 is a dummy variable equal to 1 if Y2 > T and equal to 0 otherwise. It follows that T determines the average probability of selection, or essentially the "hazard rate" of sample selection bias. Heckman uses probit analysis to estimate Y3 using **X**2 as the independent variable(s). Then the residual of this estimation is saved and added as an independent variable to subsequent equations. Computationally simpler is OLS, following Olsen's (1980) P-1 method, which Olsen showed generates essentially similar results to Heckman's probit-based hazard rate.

We used these basic techniques to estimate sample selection bias in several ways. We started with variables like those in Table A.1, that is, mayor, chief administrator, and council member survey items that varied in the number of cities from high to low. Dummy variables were created with values of 1 if data were present for the variable, and 0 if data were missing. These dummy variables were then used as dependent variables in regressions, which we may term Step 1 estimates, such as:

$$\text{V228RDUM} = a_5 + b_1\text{LPC80PIN} + b_2\text{LBLK80} + b_3\text{LPOPCHG} \tag{5}$$

$$\text{LGRPSDUM} = a_6 + b_4\text{LV799} \tag{6}$$

where

V228RDUM = 0 if data are missing for V228 and 1 if data are available; V228 is the CAO's response about the importance to the city of seeking new local revenues

LGRPSDUM = 0 if data missing for LGROUPS and 1 if data available; LGROUPS is the log of mayor's response about the importance of different organized groups

LPC80PIN = Log of percent of citizens living below poverty in 1980

LBLK80 = Log of percent blacks in the city in 1980

LPOPCHG = Log of population change from 1970 to 1980

LV799 = Log population size 1980

a = intercept

b's are estimated OLS regression coefficients

The "hazard" residuals from Step 1 Equations 5 and 6 are then saved (e.g., HZNV228P and HZNLGRP) and added as independent variables along with others for subsequent substantive analyses (Step 2), such as:

$$\text{LGROUPS} = a_7 \tag{7}$$

$+ b_i$ {
HZNV228P Residual hazard/P-1 term
LDOWNSDI log of Downsian distance of mayor
LFOG Form of government
LPARTY Log of Left/Right Continuum (V136R)
LBLK80 Population black 1980
LFPIDX84 FUNCTIONAL PERFORMANCE INDEX 1984 (LN)
LGROUPS Importance of organized groups

$$\text{LMYSPINF} = a_8 \tag{8}$$

$+ b_i$ {
HZNLGRP Residual hazard/P-1 term
LLDRSTA2 Leaders vs. staff power index 2
LPCTORG Pct. organized mun emps-1983
LGROUPS Importance of organized groups
LPARTY Log of Left/Right continuum (V136R) LJ
LBLK80 Population black 1980

Results for Equations 7 and 8 are in Table A.4. The regression coefficients for the hazard or P-1 residual terms (HZNV228P and HZNLGRP) in these Step 2 models indicate the magnitude of "sample selection bias." One also compares the coefficients of all other independent variables in the Step 2 models against those from another identical regression except that it omits the hazard (or P-1) variable, to see if any coefficients are changed by inclusion of the residual hazard variable.

Following these procedures, we estimated dozens of regressions for models with independent and dependent variables chosen to see if they might generate sample selection bias. We first examined the b coefficient of the residual hazard term, and then looked at differences in coefficients for each independent variable in the full Step 2 equation estimated with and without the hazard term. We were surprised to find remarkably few sample selection effects using this methodology.

We experimented further, using a variety of alternative specifications to the models in Equations 5 and 6, to see if we might generate some sample selection bias effects. It soon is obvious that if one includes variables at Step 1 that are associated with a dependent variable in Step 2, then one will observe significant "sample selection bias."

Conclusions on Sample Selection and Related Issues

Such experimentation led to our general conclusions, first about our data: the magnitude of sample bias is generally quite limited. Second about the methods for detecting bias: after several months of modeling alternatives, we stopped routinely including hazard rates in most regressions as some (e.g., Burk 1983) suggest, because the magnitude of the "bias" depends so heavily on (a) which variables are chosen for the Step 1 equation estimating the hazard rate and (b) how interrelated these variables are to others in Step 2. These rapidly become more standard issues of (a) substantive model specification, including (b) possible substantive differences among subgroups in the population (e.g., cities with more black residents or cities in the West, which we identified as different in how they interact with other variables in chapters above), and (c) multicollinearity, rather than (d) selection bias.

An elementary example: if population size is used to estimate the hazard rate in Step 1, and in Step 2 one analyzes variables strongly related to population size, the normal hazard rate test will indicate "considerable sample selection bias." But if instead one includes in

TABLE A.4 Two Step 2 Regressions Show Low or Insignificant Coefficients for the Residual Hazard Terms

Dependent Variable LMYSPINF = Mayor's Spending Preferences × Mayor's Influence

Multiple *R*	0.64306
R Square	0.41352
Adjusted *R* Square	0.40199
Standard Error	0.58173

	DF
Regression	6
Residual	305

Variable	*B*	*Beta*	*T*	*Sig T*
HZNLGRP	–0.29964	–0.009548	–0.204	0.8383
LLDRSTA2	0.06179	0.05492	1.241	0.2156
LPCTORG	–0.09838	–0.02804	–0.616	0.5382
LGROUPS	1.07329	0.64517	14.521	0
LPARTY	0.04892	0.02092	0.457	0.648
LBLK80	–0.28294	–0.04757	–0.977	0.3292
(Constant)		0.99376	0.243	0.8081

Dependent Variable LGROUPS = Summed activity level of major organized groups in the city

Multiple *R*	0.36302
R Square	0.13178
Adjusted *R* Square	0.11492
Standard Error	0.42541

	DF
Regression	6
Residual	309

Variable	*B*	*Beta*	*T*	*Sig T*
HZNV228P	0.64754	0.11142	1.894	0.0592
LDOWNSDI	0.62018	0.31976	6.018	0
LFOG	–0.06794	–0.05383	–0.995	0.3207
LPARTY	0.07127	0.0507	0.912	0.3624
LBLK80	0.24516	0.06857	1.202	0.2303
LFPIDX84	0.03759	0.03632	0.63	0.5293
(Constant)		0.44219	1.802	0.0725

SOURCE: Data from U.S. Fiscal Austerity and Urban Innovation Project.
NOTE: These are OLS estimates of Equations 7 and 8. Variables are defined just below Equations 7 and 8.

Steps 1 and 2 equations variables orthogonal to population size, the method will detect "no sample selection bias." This most elementary but critical point is passed over with remarkable silence in many papers on sample selection (although Stolzenberg and Relles 1993, forthcoming are exceptions that reach conclusions similar to ours). These are powerful reasons to be cautious about seeking to use the hazard or P-1 type method as an integral part of continuing substantive work, rather like a virus alert program running in a computer.

Our main concern is primarily to specify how different urban processes operate, rather than simply to describe the U.S. population of cities. We typically model these processes using regression analyses of patterns compared across cities. Sample selection bias is important in this context insofar as nonresponse patterns could lead to biased coefficients. This could occur, for example, if smaller cities had poorer residents and mayors who responded less often. But our study differs, for example, from a survey of individuals who use drugs (as in Burk 1983) where his concern was that (a) heavy drug users may not participate in the survey and (b) survey participants may differ systematically in their behavior from persons who do not respond. We do not anticipate the sort of selection bias he describes.

Further, our study suffers only partial, not complete censoring: we have data for the complete population for some variables, but less complete data for other variables. Such partial censoring makes it simple for us to specify possible selection bias more readily, more precisely, and more directly using the subset comparison method. This is not to deny the obvious elegance of the hazard rate method, or that important insights may flow from its proper use, but to suggest that it is a tool quite sensitive to specifics: a powerful microscope to reexamine certain findings.

For a general overview of potential sample selection bias, as well as distortions induced by analyzing subsets of cases created by combining variables from these data, the subset comparison method is simple but comprehensive. Applied here, its strong substantive conclusion is that sample selection bias is quite limited in the U.S. FAUI data.

Recodes

Analogous to the sample selection issue is recoding items left blank by the respondent. Many alternatives in the literature (e.g., Barnett and Lewis 1978), such as assigning blank cases the score for the sample mean, lack a clear substantive rationale. For each individual item, we sought to determine what the substantive meaning of blank was to our respondent and recoded it accordingly. This meant different recodes for different items. Two examples were FAUI survey items that asked the mayor and council members to report how active 13 participants were in local affairs and how fully the government responded to them (the full mayor's items are Q8 and Q9 in the Chapter 2 Appendix.) We experimented with alternative methods for coding blank cases, especially as (a) missing or (b) assigned a score close to the lowest score on our scale (slightly different to permit us to analyze the recoded cases separately). Quite convincing for us was examining responses for individual cities we knew and reinterviewing certain local officials. We found when we probed with the participants that items they left blank were genuinely "unimportant." We similarly scanned response patterns of blanks across the entire questionnaire for all respondents. We found virtually no cases where respondents seemed to fill in only selected items across their whole questionnaire, which they might do if they were in a hurry or sloppy—common in survey research of the general population. That we did not find such random blanks, but rather quite systematic nonresponse patterns that were substantively appropriate in the cases we checked more carefully convinced us to recode these items using procedure (b).

By contrast, the item asking the mayor about his or her spending preferences in each of 13 policy areas (Q4) was left blank quite systematically for some of the 13 items. These, we found from fiscal data for individual cities, closely matched the functions for which the city was responsible, again showing care in responding. We coded these items as missing.

Functional Performance Analysis: A Solution to Differences in Local Government Responsibilities

New York City provides about 90 percent of total local services, measured by expenditures of all local governments in the New York City area. Yet the City of Los Angeles provides only about 30 percent; the other 70 percent come from school districts, special districts, and the county government. Many participants, like Mayor Abraham Beame during the 1975 New York fiscal crisis, maintained that New York's fiscal problems were caused by its wide functional responsibilities.

Others suggested that fiscally liberal past officials expanded functions, thereby generating commitments to service that may cause strain if grants are cut. Or unions, neighborhood groups, and other participants may develop more intense activities if the local political stakes are higher, and they can concentrate on a single local government. Each of these is a distinct substantive hypothesis that past research could never precisely address.

Comparing spending and leadership across U.S. cities has long been plagued by this problem, leading many to feel that "comparisons are impossible." In *City Money,* Clark and Ferguson (1983) developed the functional performance (FP) approach as a solution. It essentially consists of creating an FP score for each city, based on whether or not the city spends money on each of some 67 subfunctions. If the city spends on one of these (above a minimal threshold), it is assigned 1; cities spending below this threshold are assigned 0 for that function. Each of the 67 subfunctions is then assigned a weight that is simply the average level of spending per capita on the function by all U.S. cities. The FP score for each city is calculated by simply multiplying 67 weights by 67 dummy variables, and summing them up:

$$FP = \sum_{i=1}^{67} F_i W_i$$

where

FP = Functional performance score for the city
Fi = 1 if the city performs the function, 0 if not
Wi = Weight for the function

This FP Index may be included as an independent variable along with others in regression equations where the dependent variable is a spending or debt measure, or any variable that may be associated with functional responsibility. This setup permits treating the FP range as a substantive variable, the effects of which on dependent variables, or intercorrelations with any other variables, may be directly analyzed. The FP approach is thus more direct and precise than such past alternatives as using common function expenditures, combining state and local expenditures (only applicable to entire state-area governments), combining all overlapping local governments (which aggregates the city with all other local governments), studying a single function such as police, and so on. The FP approach is discussed further and contrasted with these alternatives in Clark and Ferguson (1983, pp. 46-52, 314-319) and Clark, Ferguson, and Shapiro (1982). It has become widely used in municipal policy output studies.

Note

1. In specifying multiple regressions we were concerned about low degrees of freedom (DF) exacerbated by variables with low *N*s and nonoverlapping missing cases. We limited most substantive analyses to those with DF = ≥ 75.

References

Achen, Christopher. 1986. *Statistical Analysis of Quasi-Experiments.* Berkeley and Los Angeles: University of California Press.

Barnett, Vic and Toby Lewis. 1978. *Outliers in Statistical Data.* New York: John Wiley.

Blau, Peter M. and Otis Dudley Duncan. 1967. *The American Occupational Structure.* New York: John Wiley.

Bradburn, Norman M. and Seymour Sudman. 1979. *Improving Interview Methods and Questionnaire Design.* San Francisco: Jossey-Bass.

Burk, Richard A. 1983. "An Introduction to Sample Selection Bias in Sociological Data." *American Sociological Review* 48(June):386-97.

Caputo, David A. and Richard L. Cole. 1977. "City Officials and Mailed Questionnaires." *Political Methodology* 41:271-87.

Clark, Terry Nichols. 1981. "Fiscal Strain and American Cities." Pp. 137-56 in *Urban Political Economy*, edited by Kenneth Newton. London: Frances Pinter.

——— and Lorna Crowley Ferguson. 1983. *City Money: Political Processes, Fiscal Strain and Retrenchment.* New York: Columbia University Press.

——— and Robert Y. Shapiro. 1982. "Functional Performance Analysis: A New Approach to the Study of Municipal Expenditures." *Political Methodology* 8(Fall): 187-223.

Dillman, Don A. 1979. *Mail and Telephone Surveys.* New York: John Wiley.

Heckman, James J. 1979. "Sample Selection Bias as a Specification Error." *Econometrica* 45:153-61.

Kish, Leslie. 1965. *Survey Sampling.* New York: John Wiley.

Olsen, Randall. 1980. "A Least Squares Correlation for Selectivity Bias." *Econometrica* 48:1815-20.

Stolzenberg, Ross M. and Daniel A. Relles. 1993. "Correcting for Sample Selection Bias in Regression." Manuscript, University of Chicago.

———. Forthcoming. "Theory Testing in a World of Constrained Research Design." *Sociological Methods and Research.*

Index

About the Contributors

Cal Clark is Professor and Head of the Department of Political Science at Auburn University. He received his Ph.D. from the University of Illinois at Urbana-Champaign in 1973. He previously taught at New Mexico State University and the University of Wyoming and was a visiting professor at Chung Yuan Christian University in Taiwan. His recent publications include *Flexibility, Foresight, and Fortuna in Taiwan's Development* (1992) as coauthor; *The Evolving Pacific Basin* (1992) as coeditor; *Women in Taiwan's Politics* (1990) as coauthor; *Taiwan's Development* (1989) as author; *State and Development* (1988) as coeditor; and *North/South Relations* (1983) as coeditor. His work has appeared in the *American Political Science Review, Business and the Contemporary World, Comparative Political Studies, Harvard International Review, International Studies Quarterly, Jerusalem Journal of International Relations, Journal of Asian and African Studies, Journal of Conflict Resolution, Political Methodology, Political Psychology,* and *Western Political Quarterly.* He also has served as coeditor of the *Journal of Developing Societies* and as managing editor of *International Studies Notes.*

Terry Nichols Clark is Professor of Sociology and Chair of the College Sociology Program at the University of Chicago. He has published more than 25 books, including *City Money, Urban Policy Analysis,* and *Financial Management Handbook for Mayors and City Managers,* and was editor of *Research in Urban Policy* (five volumes published to date) and *Monitoring Local Governments.* He is Coordinator of the Fiscal Austerity and Urban Innovation Project, a survey

U.S. cities with populations over 25,000, with parallel surveys in 38 other countries. He holds M.A. and Ph.D. degrees from Columbia University and has taught at Columbia, Harvard, Yale, the Sorbonne, and UCLA. He has worked at the Brookings Institution, the Urban Institute, the Department of Housing and Urban Development, and the U.S. Conference of Mayors. He has consulted with many financial institutions, local governments, and fiscal analysis groups over the past 25 years.

Edward G. Goetz teaches in the Housing Program at the University of Minnesota. He has published articles on local housing and economic development policy and a book, *Shelter Burden*.

Rowan A. Miranda is Director of the Office of Management and Budget for the City of Pittsburgh. He is on leave as Assistant Professor from the University of Pittsburgh's Graduate School of Public and International Affairs. He holds a doctorate from the University of Chicago's Harris Graduate School of Public Policy Studies. His research has appeared in *Public Administration Review, Urban Affairs Quarterly, Public Productivity and Management Review,* and *Research in Public Policy*. Miranda was awarded the 1992 Best Dissertation in American Public Administration Award by the National Association of Schools in Public Affairs and Administration (NASPAA).

Oliver Walter is Dean of Liberal Arts and Professor of Political Science at the University of Wyoming. He received his Ph.D. from the University of Illinois at Urbana-Champaign in 1972. He has been on the faculty of the University of Wyoming since 1970 and served three terms as Department Head of Political Science. His primary areas of teaching and research have focused on voting behavior and municipal politics. He is editor of *Political Scientists at Work* and *Politics in the West* and has edited a special issue of the *Social Science Journal*, "Politics in the West." He has published widely in such journals as the *Journal of Politics, Public Opinion Quarterly,* and *Western Political Quarterly*.

Norman Walzer is Professor of Economics and Acting Dean of the College of Business at Western Illinois University. He holds a Ph.D. from the University of Illinois at Urbana-Champaign and specializes in state and local public finance issues. He has participated in many conferences and books of the Fiscal Austerity and Urban Innovation Project, including Poul-Erik Mouritzen, ed., *Managing Cities in Austerity* (Sage, 1992). He has written extensively on urban public finance topics, with articles in *Review of Economics and Statistics, Land Economics, Industrial and Labor Relations Review, National Tax Journal, Public Finance Quarterly, Public Finances/Finances Publiques, Journal of Community Development Society,* and many others. He has authored or edited eight books; the most recent is *Rural Community Economic Development* (1991). He also has advised numerous local governments and state agencies on economic development and public finance issues. His current research focuses on financing local infrastructure and the effectiveness of local economic development incentives on job creation.